PHILIPPA SCOTT

LUCKY ME

PHILIPPA SCOTT

LUCKY ME

The Kenilworth Press

First published in 1990
by The Kenilworth Press Limited
661 Fulham Road, London SW6 5PZ

British Library Cataloguing in Publication Data
Scott, Philippa
Lucky me.
1. Natural history. Biographies
I. Title
508. 092

ISBN 1–872082–11–4

Designed by Jayne Gorecki

Typeset in Bembo Roman 10½ on 12½
Typeset by DMD, St Clements, Oxford
Printed and bound in Great Britain by
Butler & Tanner Ltd, Frome and London

Picture Credits
All the photographs are from Lady Scott's personal collection, except for the following: Birmingham Gazette, 163; William Crooke, 15; Country Life, 146; Richard Ellis, 12; Erica, 137, 151; Keystone, 140; London News Agency, 15; Ramsey & Muspratt, 55; Sport & General, 150, 151; Wirral Photo Co., 163.

For Dafila, Falcon and Nicky

Acknowledgements

This book would never have been written without Peter's encouragement. Fortunately he had read it before he died and wanted me to have it published. Much of the material came from letters to my mother and to my brother Evelyn. I am grateful to him for his help throughout.

I would like to thank Nicola Little, who typed the manuscript, and also June White, who has helped to see the book through its final stages. Without Barbara Cooper the book would not have been published. Working with her has been fun and she has been endlessly patient with me trying to sort out my English and my innumerable snapshots and photographs. My grateful thanks to her.

Contents

Illustrations

Front cover: On board *Beatrice*, 1949
Frontispiece: Passport photograph, Reykjavik, 1951
Back cover: Wedding photograph, 1951
Chapter titles: Drawing of bee by Sir Peter Scott

Introduction

I first met my sister in a hospital in Bloemfontein in November 1918. It is my earliest memory, and I remember turning away. Jealousy? It was obviously a great event, the result of a three-hour train journey of 60 miles. With the then great age-gap of three years between us we both became loners for a while.

A farm in the Orange Free State was in many ways a perfect place for children: there were animals and birds and the wide veldt stretching to the far horizon. There were people, too — not many, but characters: Uncle Reavely and Major Tylden, Rachel Tinson, Phil's governess, Mac the farm manager and his wife Margaret, Miss Barlow the nanny, Adam the gardener, farm workers with familiar faces but forgotten names, and the occasional distinguished visitor — even the never-to-be-forgotten Prince of Wales. Ever present, of course, were our loved but sometimes shadowy parents.

Phil has the gift of remaining aged seventeen, surviving even the recent shock (to me) of discovering the truth of 'sweet seventy'. She is still special, and has retained the freshness and radiance which makes this account of her early life so readable. She was always a rebel, a trait which emerges at intervals in the book.

She is small in stature, but despite appearances is tough and adaptable: she has *had* to be, on a horse, up a mountain, or scuba diving; with Royalty or Basuto children; with 'twitchers' or stone age tribesmen; in fact with all sorts and conditions of men and women.

For a person who has strong likes and dislikes, Phil has been both loyal and discreet. There is no 'dreadful man' or 'dreadful woman' but there are plenty of 'lovely people'. Working with Peter at Edwardes Square or Slimbridge in the early days must have been great fun but can't have been easy. Her initial 'empathy' with Peter was fortunate. They were two people with so much in common — different, but utterly complementary. Lucky Phil. Lucky Peter.

Evelyn Talbot-Ponsonby
Berkhamsted
1990

Prologue

'In my beginning is my end' *(East Coker — T.S. Eliot)*

We were married on 7 August 1951 in the British Embassy in Reykjavik. The sun shone, the sky was blue, and I wore a short white dress. We had only been back from the interior of Iceland for two days. There was just one day to get all the paperwork organised, a new passport for me and a ring.

I was given away by a giant of a man — Finnur Gudmundsson, Director of the Napturagrapasapnid (Natural History Museum), who was the leader of our Pinkfooted Goose expedition, and Iceland's most distinguished ornithologist. Bryan Holt, who ran everything connected with the British Embassy, was Peter's best man; James Fisher would have been, but sadly he had to return to England the day before. We were married by Jack Greenway, who was about to retire from the Foreign Office and welcomed this opportunity to do something he had never done before. Privately afterwards he said he would give the marriage ten years.

Our guests at the party were only a handful of people: the Embassy staff and a few from the American Embassy. There was champagne and a wonderful cake and I agreed to obey. It was better than either of the two alternatives I had long ago imagined — a Registry office or Winchester Cathedral (which has the longest aisle in Britain). That night we flew north to Akureyri where we stayed before being taken next day by Jeep to Lake Myvatn. There we had a week's holiday listening to the Long Tailed Ducks' lovely call, and trying unsuccessfully to catch Harlequin Ducks for Slimbridge. Now I was Mrs Peter Scott.

Why did this happen to me? Let's go back to the beginning. Few of us know much about the early lives of our parents — how they thought and how they felt. We know them only as old people, and cannot imagine them young, insecure and full of hope. These 'Memoirs' are written for my children and perhaps my grandchildren.

My father, Commander Frederick Talbot-Ponsonby, in 1911.

Home in South Africa

Peter wrote in his autobiography *The Eye of the Wind* that he was the happiest man he knew. I think I must be the luckiest person I know. Though life may not have been a bed of roses all the way, I can honestly say that more marvellous things have happened to me than to most people. I am a pessimist and well aware that many awful things could still happen to bring me my share of bad luck. But nothing can ever take away the good luck, the happiness and the adventure I have had in life.

Whatever comes,
One hour was sunlit,
And the most high Gods
May not make boast
Of any better thing
Than to have watched
That hour as it passed.

I have not been able to identify this quotation, but I like it.

Since heredity plays some part in one's life, I must go back to the beginning and start with my parentage. My paternal grandfather was the first Talbot-Ponsonby and 'squire' of the village of Langrish near Petersfield in Hampshire. His father was Admiral Sir Charles Talbot and his mother, Charlotte, was the daughter of Major-General Sir William Ponsonby who distinguished himself and was killed at the battle of Waterloo (and to whom there is a monument in the crypt of St Paul's Cathedral). On the death of his mother, my grandfather assumed by Royal Sign Manual the name and arms of Ponsonby in addition to those of Talbot in accordance with the will of his uncle who died childless. My grandfather inherited estates in Ireland, which brought little but trouble and ended in a hearing where the mass of evidence makes both absentee landlord and impoverished tenants objects of sympathy.

The estate of Langrish was inherited through my grandmother,

Constance Delme-Radcliffe, whose uncle John Waddington left it to her. More about Langrish later.

My father, born in 1879, was one of eight children (six boys and two girls, though the first boy died aged two). He joined the Royal Navy at an early age, following the family tradition to the third generation. He achieved the rank of Commander but was invalided out during World War I, only to return again to serve in Naval Intelligence. His health finally forced him to retire before the end of the war. At that time there was no cure for tuberculosis of the lungs, and a high, dry altitude was recommended. The Orange Free State in South Africa was ideal. Here at 5,000 feet above sea level the summers were hot and the winters cool and dry. With his country background, farming was in his blood and he bought a farm on the Westminster estate 60 miles east of Bloemfontein, not far from what is now Lesotho. Farms on the Westminster estate were all called after places near the Duke of Westminster's family seat in Cheshire such as Halkin, Eaton and Saighton. The farm my father bought had the name Beauchamp.

My mother was born in 1881, and was the tenth and youngest child of John Ritchie Findlay who is described in *The Encyclopaedia Britannica* as a philanthropist. He owned *The Scotsman* newspaper and presented to the nation the Scottish National Portrait Gallery, opened in Edinburgh in 1889. The family seat, Aberlour House, was an enormous mansion close to the river Spey at Aberlour in Banffshire. It is now the preparatory school for Gordonstoun. The Findlays also owned two houses in Rothesay Terrace in Edinburgh. My mother with her long golden hair was a great beauty in her youth, but did not get married until she was thirty-three — perhaps because she was the youngest of seven sisters. I know nothing of her early life, but doubt if she had as interesting a time as I did before I was married at the age of thirty-two. By the time I came on the scene, her eldest brother, John, by now a baronet, was running the family business and estates. I remember him as a very handsome, upright, friendly man, and he was much liked and respected locally, as was his wife, Aunt Hetty, whose twin brother Roger Backhouse married one of my mother's older sisters, Dora. Aunt Hetty (Harriet Findlay) was created a Dame in her own right for services rendered to innumerable charities.

My mother's early married years were spent at Englefield Green in Surrey, but after my father bought the farm in South Africa she joined him in 1917, taking with her my brother Evelyn, aged two. It

My mother, Hannah Findlay, at the time of her marriage in 1914.

My father at the races in 1928.

cannot have been a pleasant wartime journey from Southampton to Capetown: four weeks in a blacked-out ship with all the water-tight doors shut, especially as she was always seasick. From Capetown there was a two-day train journey to Westminster on the high veldt.

The farm homestead, although built of stone with a pleasantly tiled roof, was fairly primitive. It is difficult for me to imagine how my mother, with her sheltered background and quiet, gentle manner, could have coped with the situation — the day-to-day running of the household, the many employees, and the development plans — but there were undoubtedly inner depths to her character.

The house was enlarged, a Scottish farm manager was employed and his wife engaged as cook. One of my father's most passionate interests was gardening: not so much the digging and sowing, but the landscaping and designing. And so the garden grew and grew — hedges, pergolas, terraces, rose beds, rockeries, vines, violets in falling terraces between tall pine trees and candelabra cacti. There were long herbaceous borders, lawns which required endless watering, a huge asparagus bed and a vegetable garden. All this in an area of frequent and lasting droughts.

Even until the last year of his life my father was still planning further developments. His head gardener, a lovely Basuto man, named Adam by my father, was a great friend of mine — only occasionally was I in trouble with him. My happiest memories of him were being taken into his large mud-walled, thatched-roofed potting shed, which had the glorious smell of all good potting sheds, and of drinking cool, clear water out of one of the corrugated iron rain tanks — from his felt hat. It was indeed a privilege — but, alas, one day I was caught by my mother and this delight was forbidden from then on.

The farm can scarcely have provided sufficient profit for all these projects — the huge and beautiful garden, the nursery wing, the kitchen at the back of the house, the rondavel study on the end of the new wing, and the necessary servants and garden boys. I suspect my mother must have contributed considerable funds, which may be why I was brought up to believe that a girl's money should always be guarded by Trustees, so that her husband could not squander her fortune. But to suggest that my father squandered his money would be unfair. There was one major financial disaster in the early years in South Africa when he was persuaded to invest in a plantation at George in the Cape Province. The matter was not openly discussed in front of the children, but he was evidently badly let down by someone and the scheme collapsed.

If not highly profitable, at least the estate was successful. It was a mixed farm, growing maize and barley. There were sheep, beef cattle and a dairy herd, of which my father was justly proud. Bulls were always returning from the annual show in Bloemfontein bearing prize rosettes, and I still have medals which were won at the show.

My father was elected President of the Shorthorn Society of South Africa and visited Argentina officially in this capacity. A striking-looking man of middle height with mutton chop whiskers, bright blue eyes and a wide smile, he had a magic personality and since his retirement from the Navy had always been known as 'The Commodore'. He was not very interested in his offspring — but those were the days when children were to be seen and not heard. Although I was rather overawed by his presence, I was always aware of him as someone rather special. It seemed that our house was the centre of social activity in the neighbourhood. He loved giving dinner parties, for which he had a fund of good stories.

Racing was a recreation which he could only indulge in on visits to England. At one time he shared a race horse called 'Rear Admiral' with an old flame of his living near Wolverhampton. The horse was evidently moderately successful, as his portrait was painted, and a treasured shoe still hangs in my house. He was a friend of Rudyard Kipling's, whom he greatly admired, and was known also to many distinguished people in South Africa who visited him at Beauchamp, including General Smuts and Sir Lionel Phillips (after whom I was named).

From the time I was born in Bloemfontein in the middle of the Spanish flu epidemic, until I was five, I had a succession of nannies from England, only one of whom I remember. She was called Miss Barlow, and she is memorable mainly for her massive false bun and her nightly administration of syrup of figs. I have never been able to eat figs of any kind since.

My earliest memories are of ponies — of yelling when put up in the saddle in front of my father on a Basuto pony called John Willie who had served in the Boer War, and my father's evident disgust at my reaction. Toddles, as I was called by him then, was in disgrace. At

Left: Me in South Africa, aged 3½.
Below: In the farmyard at Beauchamp with my Basuto pony, Billy, in 1928.
Right: With Tigger and my brother Evelyn at Beauchamp in 1926, when he was 11 and I was 8.
Far right: With Scot just outside the garden at Beauchamp in 1927.
Below right: My friend Mahlasnyan and my grey pony, Scot.

the age of three I remember falling off my brother's little black pony, Piccanin, who bolted through the farm gate as my mother was closing it. This time there were clouds of glory not shame, for not making too much fuss. My own first pony was a bay called Billy, and there are photographs of me looking somewhat ant-like on his back.

Ponies were a great feature of our life at Beauchamp. My brother rode to his private lessons four miles away with another child, Jackie Sayer, escorted by a Basuto boy.

Three miles away — or was it only two; distances seemed so much greater then — was the farm called Saighton, where one of my mother's favourite sisters, Aunt Florence, came to live with her husband, Uncle Reavely (Maitland). Uncle Reavely had two beautiful daughters called Arundel and Cicely, but both were quite a bit older than us. The Maitlands had a farm manager and were not at Saighton all the year, but when they were, Arundel and Cicely were always very pleased to play with their younger cousins, and the parents were in and out of each others' houses all the time. Uncle Reavely was a larger-than-life character with a drooping grey moustache. He always drank his tea out of the saucer, and had an infectious sense of humour and a very loud voice, which could be quite terrifying. Aunt Florence was a quiet, gentle person, almost saintly, who had a great affinity with the young. Aunts Florence and Dora were, with my mother, the three youngest of the Findlay girls and remained close to each other all their lives.

Our chief means of transport was a four-wheeled buggy drawn by two flea-bitten greys called Peter and Paul. But my mother and I rode all over the farm on small, wiry, unshod Basuto ponies. When I was seven I was given a pony for my birthday and was allowed to choose from three, brought from Lesotho for us to see. In spite of the fact that I chose the grey because of his colour and because he pranced with an arched neck, he turned out to be a great success and I named him Scot. My father, however, decided to buy the bay which he liked better, which was called Moutla (meaning 'little hare'). Moutla then became the mount of my seventeen year-old Basuto companion Mahlasanyan, who came riding with me and from whom I learnt a smattering of the Sesuto language.

It was around this time that the then Prince of Wales, later King Edward VIII, visited Westminster and came to call on my father. We all lined up on the stone steps leading to the *stoep* (verandah) and front door of the house. The night before, Miss Barlow had put my hair in paper curlers. My mother was furious and made her brush out my wispy blond hair with a damp brush. I was supposed to present the Prince with a bunch of home-grown violets to put in his button hole, but when it came to the point I held it firmly behind my back and he didn't receive it. Again I was in disgrace, and long remembered the shame. Later the Prince of Wales was to return to Westminster when my father was very ill, and although he came to the house and I was presented to him, no buttonhole was offered, but I developed a romantic image of this handsome young man and I continued to admire him for many years. A ball was given locally in his honour and I was furious that I was too young to be invited.

By 1923 my mother seemed wholly pre-occupied with my father's indifferent health and with helping to run the house, the garden and to a certain extent the farm. A governess, Miss Tinson, was imported from England and she stayed with me until we finally left South Africa in 1931.

She brought a whole new dimension into my life — a warm, lasting affection, discipline and, apart from French (which was often with tears), she managed to make my lessons fun as well as interesting. Her holidays were spent with friends of my parents on a farm on the border of Lesotho. Later she was to marry the farmer, Major Tylden. It was he who taught me to ride, and he always brought laughter when he came to the house. His youngest daughter, Betty, who was about my age, was held up by Miss Tinson as a shining example of competitive effort. She was taught at home by her parents, and although she and I got on very well together I was somehow made to feel intellectually inferior: a feeling with which I am now still quite familiar among certain types of people.

Because of my father's TB the family often avoided the Southern Hemisphere winters by visiting England during the summer. By the time I left South Africa I had completed the three-week voyage by Union Castle line from Capetown to Southampton or vice versa fifteen times, beginning at the age of three.

On the very early voyages we were accompanied by a nanny, and once by the farm manager's wife — a splendid Scottish lady who spoiled me. (I used to visit her cottage near our house to be plied with fresh Scotch pancakes with farm butter, and other goodies.) The journeys by sea were wonderful holidays for a child. My mother was always seasick, and I knew the ships so well that it was like being in a familiar hotel. That I behaved very badly is certain. In those days shyness — at any rate with my contemporaries — had not hit me and I used to join a gang and run riot. I remember climbing through the window of the lounge and tearing a huge rent in a cotton dress sent to me by my grandmother. The scolding from my mother was nothing, because I hated the dress and knew with a lift in my heart that it was really unmendable.

By the time that I was ten I had become quite accomplished at the deck games, and I embarrassed the authorities by winning the adult championships in deck tennis as well as trophies in the deck quoits. I still have a number of cups, spoons, table napkin rings, etc, emblazoned with the Union Castle crest.

The wildest time of all perhaps was when we travelled at the same time as a contingent of boy scouts, who were on their way to a World Jamboree in England. As an extremely well-developed twelve year-old it was my first lesson in attracting the opposite sex, but I was soon in trouble for spending my time in the Third Class with them. Once, too — perhaps on the last voyage in 1931 — I decided that I had fallen violently in love with the handsome blond swimming instructor, who occupied my thoughts for some time after that.

But the best part of going to England was coming back home — back to Beauchamp. Even the railway station at Westminster had a magic quality: the board on the low platform giving the height above sea level, the neat hedges, the small red-roofed building, the tall grain silos across the railway lines, and a little further beyond the bare yard, the store. The store was wonderful: hung with brightly coloured blankets, which were worn by the Basutos working in this area, and with all the other useful things to be seen in general stores at that time in a rural district. And it had a smell not unlike that of potting sheds.

But anticipation built up before the station was even reached. The railway line passed close to the far end of the farm boundary and there was an unusual view of the *kopje* under which the house nestled. Excitement had to be controlled — emotion was never shown in the family — but that didn't matter: it was there, secret, glowing — and almost bursting within.

In the early days we were met by the buggy and Peter and Paul the ponies, with the Scottish farm manager, Mr McKiddie, then later by the bulbous-looking Dodge motor car. A rough track, back over the railway lines, past the church-cum-school, the long straight earth track by the line of gum trees, a right turn, down and up through the *donga* (a deep ravine normally dry) past the doctor's house in the trees and finally to the little native village with its mud huts and aloe plants, where all the African workers on our farm lived (it was known as 'the location'). Then the stone gate-way, the quarter-of-a-mile drive lined with mimosa trees on one side — which in spring filled the air with their strong, sweet scent — and an orchard with peaches, apricots and pears on the other. The drive was almost impassable in a motor car after rain. The earth could turn into deep black mud, and there were occasions when the Dodge got stuck, which was exciting for a child and often required a team of oxen to heave it out. Peter and Paul managed rather better. But rain was rare and drought more common.

At last there was the house and garden and a warm welcome from our farm helpers and the dogs. The house stood back from the front gate, which was flanked by tall cypress trees. A flagstone path led up across the terraced lawns with flower beds to the old part of the house which was built of the warm yellow local sandstone, single storey with a red tiled roof and a *stoep* along half of the front. To the right and further back was the newer nursery wing which was up some steps and half hidden by a hedge. There were two large rooms with a pink-tiled *stoep* all along the front. The first room next to, but not actually joined to, the house, was the nursery. Beyond was a large spare bedroom and at the end of the *stoep* was my father's study, rondavel-shaped but built in stone. This wing was all roofed with red-painted corrugated iron. When it hailed, which was not infrequently, the noise was tremendous. A bokmakierie — a beautiful black and yellow shrike with a loud song — lived in the flower bed below the *stoep*.

Of all the magic places on the farm the *kopje* was perhaps the most special. The house and farmyard with its shady gum trees, the pine trees in the garden, the neatly-clipped hedges, the stone terraces, the fountain, the little orchard across the drive, the stack yard and cow sheds and kraals all lay at the foot of the lowest part of the *kopje*. A farm boundary corner was at the highest point on the left. Here it was rocky and barren with sparse vegetation, but below this point was a special playground with a haunted atmosphere. Shallow caves in the white sandstone rocks were surrounded by strongly scented bushes, and there was a rich, warm smell of aromatic plants. It had once been the home of Africans, but only broken-down walls remained.

We occasionally found snakes on the *kopje* — the poisonous Ringkols being the most dangerous — but we kept our eyes open for

Left: Miss Tin sitting in the garden at Beauchamp.
Below: Westminster, the station for Beauchamp. I took this photograph in 1963, but the scene was the same as I remembered it from my childhood.
Right: The front of the house at Beauchamp. The garden was my father's pride and joy, and in the summer was a riot of colour.
Below right: A view of the farm and Taba Pachua, taken from the kopje.

them and never came to any harm from one. Once, Miss Tinson (known always as Miss Tin) and I found an Eagle Owl's nest with young in it, and the parent bird attacked us. Miss Tin put up an umbrella and we retreated unscathed. This made a lasting impression on me — perhaps more because of the unlikely umbrella than the angry owl. Some areas near the top of the *kopje* were covered in beautiful shiny, soft grey elephant moss. Another fascinating area had large grey slabby rocks, pock-marked with shallow rounded holes which held water after rain. It was lovely to put one's fingers in the holes and feel the smoothness. Years later my brother and I, on separate occasions, went back to the spot and traced these same holes again.

Further along the lower part of the *kopje* was another ancient site known as 'Maria's garden'. More scented bushes here, and lush vegetation, and the rocks were the home of 'dassies' (rock hyraxes) which were very shy and disappeared quickly into their holes as we came near.

South African flowers are well-known for their diversity and beauty. Dry as this area was, nevertheless a great variety were on show at different times, and it was Miss Tin who taught me their names when she was able to identify them. On the *kopje* some of the most intriguing were the *stipeleas*: low-growing, grey-leaved, with green and brown spotted, star-shaped petals. They had a nasty smell and were not at all beautiful — but they were very unusual and I found them fascinating.

Over the top of the *kopje* and down the other side was another special place to which we sometimes made expeditions for picnics. Here there was a water-filled dam with a huge old weeping willow overhanging it at the lower end. Half-way up the willow was the nest of a Hammerkop, which takes its name from its hammer-shaped head. As it is brown all over, with a yellow beak, it is not especially beautiful: yet it is a very strange creature, the size of a small egret and related to the stork family. It builds a vast nest of twigs which it enters through a hole. As I recollect, the nest at One Tree was a traditional one and occupied nearly every year. They are rather shy birds and we never had a close look at the nest, as we were forbidden to disturb it. In my mind it somehow acquired a mystical quality, rather like the fabulous Roc. It still remains one of my favourite African birds.

My first real interest in birds was inspired by Miss Tin. As well as riding, we often went for long walks. There were several belts or plantations of trees on the farm, and plenty of colourful birds. Most spectacular were the weavers and bishop birds with their hanging nests and the incredible sound of their chatter in the nesting season. Along the fences the remains of the Butcher Birds' quarry were impaled on the barbed wire. Among other favourites were the Widow Birds with their tremendously long tails, and the large Secretary Birds, which were quite common in the mealie (maize) fields. Small flocks of guinea fowl could be seen on the farm, but my father had acquired some semi-domesticated ones as well, which lived round about the back of the house near the *kopje*. One in particular became his favourite. It had large patches of white on its breast, distinguishing it quite easily from the others. My father's delight was to leave the dining room door open and watch 'Pedro' come in and eat the flies off the floor-to-ceiling gauze windows. I had the impression that my mother was not so pleased, but as children my brother and I loved all the guinea fowl and gave names to those which came near the house.

Below the large farmyard with its cow byre, stalls for the bulls, dairy, mud-walled cattle kraals and other sheds was the fenced-in stack yard. This had blue-gum trees on two sides, a muddy dam with ducks on the third side, and hen houses down at the bottom, with free-range hens which laid prolifically all over the place. The straw and mealie crops were brought in on farm carts pulled by teams of oxen — sometimes ten pairs at a time. Watching them being handled by the competent Basuto drivers was a never-ending delight. It involved much shouting and cracking of whips.

From the edge of the farmyard, in front of the house, which faced south, and beyond the little orchard on the other side of the drive, stretched out a vast plain. The brown grass of the veldt with the fenced-in camps (or fields) for stock was broken by the occasional oasis of a tree-girt farmstead. At the edge of the plain straight ahead was an isolated mountain, flat-topped with a rocky cliff below its table top. Nearly always blue, sometimes purple, it dominated the landscape. It was called 'Taba Pachua', and beckoned like a friend. Once, my mother, Miss Tin, and I made an expedition to climb it. It was a wonderful adventure and may have been responsible for my love of mountaineering. The rocky cliff was a scramble and had large aromatic shrubs growing out of the crevices. I have since learned that it was only twelve miles from Beauchamp.

Looking out from the *kopje* and further round to the left — east of the farm — were more mountains which were the beginning of Basutoland. Here the shapes were different and the hills were not so high, but they had romantic names like M'bula and Seta's Kop. Beyond the hills lay the Tyldens' farm by the Caledon River, which forms the border between the Orange Free State and Lesotho.

Most of my free time out of lesson hours was spent either in the farmyard, climbing the straw stacks, riding, or playing with the three Irish terriers which lived in the house.

At the far end of the farm a large apple orchard surrounded by a windbreak of fir trees on all four sides was a good place for riding. And we used to swim in the oblong stone tank next to the shed where the sulphur-dried apple rings were sliced and packed. The orchard was formed as a separate company — Imokilly Orchard Ltd — and was, I believe, one of the first to export apples from South Africa. This was where I learnt to swim and won *The Jungle Book* as a prize. The water was green and turgid, but that was not important. My brother Evelyn and I used to enjoy watching the apple rings being made, though the smell was horrific. It was interesting to learn the names of the successful cropping apples — Rome Beauty and White Winter Pearmain being the favourites.

Occasionally, I had friends of my own age from another farm who came for the day. Only three girls, one of whom was Jackie Sayer, Evelyn's schoolmate, came with any regularity. The son of friends of my parents whom I disliked was taken by me into the greenhouse and given red chillies, which I told him were a new and delicious kind of small red banana. I don't remember any other malicious acts on my part. My father taught me to shoot with a .22 rifle and gave me my own for my tenth birthday. With the help of my governess, I made targets by cutting out animals from children's books and setting up a simple device with strings in a strip of gum trees, so that I had moving targets for practice. My father used to give parties at which rifle-shooting competitions were held, and the prize for splitting a playing card sideways-on in a cleft stick at 10 yards was a bottle of whisky. For winning this I used to get half a crown, but his pleasure in my success meant more to me than the money — after all there was nothing much to spend it on there. Doves were very plentiful in the garden and I regret to say that I used to shoot them as they sat in the tops of the tall fir trees.

The most special and memorable event on the farm was Christmas Day. Of course it started with a stocking. Then we went to church.

The next most important happening was when the entire farm labour force from the native 'location' trooped down the long drive with their wives and children, the men wearing their lovely brightly coloured blankets, the women with their full skirts swinging, bright scarves over their heads and their babies on their backs, all singing and shouting. Out in the farmyard my father presented all the men with a 'Christmas Box', and then wrapped sweets were thrown in handfuls for the children. A great spirit of happiness prevailed — a real rejoicing.

In the garden, trestle tables were put out on the lawn under the line of fir trees and covered with white damask table cloths. As far as I know, everyone on the entire Westminster estate came to the Christmas cold buffet lunch party. Often it was the hottest day of the year. Once I remember it being 110°F in the shade. There was champagne and plenty of delicious food. An idyllic kind of Christmas Day.

After my father died and we moved to England, Christmas still had its ritual — church, presents, turkey and plum pudding — but it was a cold, grey affair compared to the bright golden happiness of a champagne party on the lawn at Beauchamp.

The first major change in my life occurred in 1923 when my brother Evelyn was left behind in England at the age of eight to go to preparatory school. He was based with Aunt Dora and Uncle Roger Backhouse, my mother's sister's family who lived in Havant. There were six children in the Backhouse family, so he was to have plenty of companionship. Aunt Dora and Uncle John were the two Findlays who married brother and sister. Uncle Roger was later to become First Sea Lord. I found him very frightening, but Aunt Dora, one of my mother's favourite sisters, was friendly and kind.

Evelyn's departure to prep school took place only a few months before Miss Tinson arrived at Beauchamp. Until then Evelyn and I had only been moderately good friends. Most of our nannies had favoured one or the other, which didn't help. 'Deliver us from Evelyn' was apparently part of my nightly Lord's Prayer. A three-year gap in those early years makes a lot of difference and Evelyn had a South African school friend in Jackie Sayer, while I had none. She was later to become a friend of mine, though she was perhaps two years older.

Once a month we went to a Church of England service in the stone-built school two miles away, on the other side of the *kopje*. I enjoyed the hymns — but I never heard any other music and, as I had no natural ear, I could only sing out of tune. In fact, there was no music at all in our home: something which seems strange to me now. We did, of course, have marvellous birds singing in the garden — sounds which, like smells, are very nostalgic.

At some period in those early years I went through a brief period of intense religious fervour, and was convinced that Jesus Christ would turn up somewhere in the surrounding veldt at any moment. What did turn up with unfailing regularity every year was a plague of Emperor Moth caterpillars on the huge weeping willow tree at the back of the house. Having stripped the tree completely they fell on the ground below, four inches long and as thick as your finger. Imagine running out with bare feet and suddenly you are among this soft, squishy yellow mass! I have never been very fond of caterpillars since then.

The most wonderful thing that happened in the whole of my childhood was the arrival of my brother Evelyn for a six months' visit in 1929. He was thirteen, I was ten. This was real happiness: a constant companion, to share my lessons and my play. I glowed with the knowledge that his maths were at a lower standard than mine, and loved him for being interested in not only all the dogs and cats, but also all my toy animals. They were a miscellaneous collection, the favourites being a teddy bear and a black cat. A row of about twelve of them was solemnly photographed in the garden, assembled for the wedding of the cat and the bear — Mr and Mrs Peter.

A few weeks later Peter was knighted and they became Sir Peter and Lady Peter. Lady Peter acquired a large wardrobe of clothes — dresses, pants, coats, pyjamas and then, of course, bedclothes — all made by Miss Tin and me. On one of my visits to England I entrusted Lady Peter to our friend Jackie Sayer. Years later Jackie visited me at Slimbridge and asked to see Lady Peter, who by that time was rather worn, but still well-equipped, having had a prolonged happy life with my daughter Dafila. Some teddy bears are very long-lived and lucky. It seems strange looking back that my pony was called Scot (albeit with one T) and my teddy bear called Lady Peter.

The only sadness during Evelyn's visit was the rift between me and Miss Tin. Led astray a little perhaps by him, I began to resent the discipline. I suppose I was a born rebel, and it was easy to make use of an older and adored brother to flout her authority, which until then had been absolute. The problems in her own private life may also have had some effect on our relationship, which remained slightly strained until the time I left the country. She then stayed in South Africa to marry Major Tylden, the father of my intellectual rival. Many years later Evelyn and Betty Tylden were to meet again when both were up at Cambridge at the same time. But looking back on that period of strained relations I realise the extent of the regret I had even at that time about the situation. I think I was rather jealous of Miss Tin's relations with the other family, but she was one of the few people I have really loved in my life. She gave me far more affection than my parents had ever done. For years after I came to England I dreamed of her — and always she was going away and leaving me. I think I was jealous, too, when she produced a daughter of her own. Not until her daughter was grown up and I visited her and Major Tylden at their home in England did I feel I had made my peace with her and had been accepted. From then on I stopped having dreams of being left by her. She could never have known how much I missed her.

CHAPTER TWO

Meeting the Relations

On our almost annual visits to Britain in the English summer we were joined by Evelyn. These visits started at my grandparents' home at Langrish House near Petersfield. I cannot remember Southampton being anything other than cold and grey, and so it set the scene. In the early years we were driven to the house by a brougham drawn by two mules. My grandmother wore a black dress and a black velvet ribbon round her neck. She had sleek white hair piled in a coil on top of her head. My grandfather had mutton chop whiskers like an older version of my father. One of my father's unmarried sisters, Aunt Brucie, lived there, too.

The grandparents and the other Aunts and Uncles who came and went were all very awe-inspiring — all except Aunt Brucie who had a herb garden in the village and seemed at least to have a sense of humour and was a little less stern. My South African accent was made fun of. 'How now brown cow' they kept saying to me. One of my aunts-by-marriage declared that I was a 'rude little girl'. Altogether, staying at Langrish was a painful affair. Evelyn was nice to me when we were on our own and we shared a common dislike for one of our temporary nannies, which served to cement our friendship further. She was Scottish, and highly superstitious. It was she who told us that we should always put on the right shoe first. Not long ago I discovered that Evelyn still obeys this rule, while I still do the opposite.

Langrish had a long, dark passage in the middle of which was a swooshing green baize door. In order to reach our bedrooms, which were opposite each other, we had to pass along this sinister passage. I am sure that the house was haunted and this passage was the nastiest place in it. On one occasion Evelyn was put in a bedroom close to the end of the passage, and he disliked it so much that he refused to sleep in it again. It had been the room of my Aunt Evelyn, who died in 1915.

There was a huge garden which at least gave us some scope for play, but it was all very different from our home in Africa. Woods, tall dark trees, laurel and yew closed in on the house, and the shrubbery smelled dark and sour. The daisies on the lawn were the nicest thing for a native of South Africa, and Evelyn made me daisy chains. There was a walled terrace below the lawn in front of the house with a rose bed beneath it. I think it must have been about 4′6″ high — much higher than me anyway — and Evelyn challenged me to jump down and over the 3ft rose bed — a monumental jump for someone as small as I was. It was important to be able to accomplish *any* of his challenges, but this was one of the most formidable which I remember.

We enjoyed playing together, but when my cousin Alathea joined us from their house a few miles away it was a case of 'two's company three's none'. She was two years older than Evelyn — an attractive, vivacious tomboy of a girl. Once, they left me at the top of a high haystack in a barn and took the ladder away. Another time we played sardines with some other children and they left me for what seemed hours searching vainly round a strange garden for the rest of the party. For me at that time Alathea's one saving grace was that she had a piebald pony called Hyde Park which I was occasionally allowed to ride. He was a marvellous show pony and hunter. Over his stable hung a mass of rosettes.

Alathea was the only child of Arthur Talbot-Ponsonby, my father's youngest brother. He was known as Uncle Noah because as a small boy he used to answer 'No-ah'. Like my father he was a good story-teller, witty and amusing, with all the famous Talbot-Ponsonby charm. Like my father, too, he loved good horses, good cattle, good wine and good company. He and my father used to go racing together. He was a farmer and was later to become land agent (or Commissioner) and friend to the late Bernard, Duke of Norfolk at Arundel.

Almost the worst event that took place at Langrish was the weekly visit to church on Sundays. Sometimes it was a sombre drive with the brougham or landau and mules. Sometimes we walked. That was better — along the cinder path with the sloping orchard on the right with the huge mulberry tree and Uncle Char's Barheaded Geese to give some diversion from the dreaded prospect ahead. We sat in a pew near the altar steps facing south. A door boxed us in. Dressing up in all those hateful clothes was bad enough. Sitting through a dreary service was worse, and for some reason putting money in the

Above: With Aunt Brucie's West Highland terrier, Hamish, at Langrish. I was five.
Left and right: My first animal pictures, taken with a box Brownie. In chronological order: Pedro, the tame guinea fowl, Beauchamp. Geese at Langrish. Mute swans at Langrish. My favourite pair of guinea pigs, at Pineridge.

collection was the ultimate in frightening rituals. I am not sure why this was so, but suspect that it was because I was caught unawares so often, and once lost my sixpenny bit.

Visiting the Scottish relations was much more fun. Three of my mother's sisters lived in a house called Kinermony about two miles upstream from Aberlour on the river Spey. The three-storey white-washed house stood on a bluff above and back from the river, with the sound of the rushing water clearly audible from the rooms on the river side. Wild raspberries grew on the slope below the house, and in front was a grass lozenge with two trimmed yew trees like green candle-flames, surrounded by the smooth pink gravel drive where I learned to ride a bicycle. A huge oval walled garden fell steeply down to a central dividing burn then rose just as steeply on the far side. The icy waters of the gurgling burn flashed through a riot of mimulus plants — orange and yellow — and ended in a shallow, round concrete pool by the wall where there was a sluice. Paddling in the pool was permitted, but damming the burn was not. Beyond the wall the burn joined another one and rushed on down to the Spey. By the back door of the house and opposite Evelyn's bedroom window was a large tree with an owl's nest in it. Just beyond were Aunt Grace's ducks which I used to help her to feed.

Kinermony was for August when the school holidays had started, and some of the cousins with whom Evelyn stayed were also fellow guests. Our contemporaries in the Backhouse family were Barbara (known as Barbe) and Joan, and their mother Aunt Dora came with them. Here we were not burdened by a nanny. An elderly lady called 'Cranston', who was a sort of housekeeper/seamstress, kept a vague eye on us: which mainly consisted of overseeing our early suppers in the 'nursery'.

Otherwise we had a wonderfully free time. None of the three Aunts was married. The eldest was Aunt Jeannie. She ruled the roost and kept the peace among the three. A quiet, gentle person with immense dignity she commanded great respect. To me they all have colours. Aunt Jeannie was dark green — this applies to tweeds and jumpers, not her complexion. Aunt Grace was also green, but a brighter green, with dark red hair. As the youngest, and so keeper of the domestic ducks, Aunt Grace was my favourite, but then she had stayed with us once at Beauchamp and knew about the things that mattered to me like ducks and dogs. Aunt Elizabeth was blue, and apart from her tweeds, her eyes were blue too. She must have been a ravishing beauty in her youth. A bit 'fluttery', she giggled a lot and was somehow less in tune with the young nephew and nieces — or perhaps it was just that here again I was considered a bit different from the others because of my South African background. With some regret I remember being very rude to her when she tried to bath me one night when my mother was out or away.

There were only two bad things which we had to put up with at Kinermony. First we were often sent on drives in the huge Daimler, six or seven of us at a time, to see the countryside. I was usually one of the two people sitting behind the glass partition with my back to the driver. I was never actually sick but always felt very sick. The chauffeur, Turnbull, was a delightful man, a great ally in our bicycling escapades, and it was usually Evelyn and Barbara, the older two, who sat up front with him.

The drives were more enjoyable when there was an objective, such as an ancient castle to visit. We must have seen every ruined castle within a radius of fifty miles. Some were favourites, and we visited them several times, scrambling over the ruined walls and poking around in every corner. Once, we went to the Culbin Sands where an entire village was buried under the sand dunes and picked up pieces of pottery and other human artifacts which were then stored in Evelyn's museum.

The other unpleasant ritual was the duty chore of fruit-picking — red currants and white currants — they were not even very nice to eat. Dyce, the dour gardener, was not our friend. He didn't approve of us and became something of an ogre. There were lovely big, sweet, juicy raspberries in the garden, too, but we were not trusted to pick those. With so much fresh garden produce, the food in the house was memorably good. I learned to hate porridge and to love pheasant and grouse. The servants, unlike the ones at Langrish, seemed to enjoy spoiling us and we used to go down to the pantry in our pyjamas to feast on the remains of what came out from the dining room at night, in spite of having had an excellent supper.

There were plenty of small, pock-marked, puddle-strewn minor roads almost entirely traffic free, lined with a riot of harebells and other lovely wild flowers, which we could explore on our bicycles. It gave us a delightful sense of freedom. One of our favourite haunts was the site of an old sawdust mill on Tom na Bent, a hill about one and a half miles away. A fast-flowing burn rushed down from the peat and heather above, so we were able to make dams and pools and

Above: Family group at Kinermony. Front *(l to r) Evelyn, Arundel and Cicely.* Centre *(l to r) Aunt Florence, my mother, Aunt Grace.* Back *(l to r) Aunts Elizabeth and Jeannie.*
Left: My grandparents' drawing room at Langrish as I remember it.
Below Left: Kinermony. The front of the house, taken from 'the lozenge'.

floods. Like a magnified sand pit with running water laid on, it was an ideal playground. In a quarry just across the road we dug out caves, one for Evelyn and Barbara, and one for Joan and me. Barbara was two years older than me, and Joan one year younger. The four of us split up quite happily into two, though sometimes we did things together.

The Scottish summer holidays continued well into our teens, and as we grew older we enjoyed them more and more, with childish games progressing to hill climbing and adventures further afield. A yearly climb to the top of Ben Rinnes, from the distillery at the foot, was a highlight. We would stop on the way to throw a penny in the wishing pool among the rocks half way along its elongated skyline. And if we weren't on our bicycles exploring the countryside there was always the Spey. There were little sandy bays in some places and there were steep tree-clad slopes to scramble along near Craigellachie, and even a railway tunnel to penetrate for the very daring.

One year, Barbara and Evelyn set off on their bicycles for Tomintoul, the highest village in Scotland, a distance of some 20 miles. Joan (or 'wee Joanie' as she was affectionately called by some Scottish relatives) and I knew all about the plan, but had been sworn to secrecy. It was only when they were not back by supper time that everyone began to worry. The concern must have been considerable for me to remember it — or perhaps I was worried about keeping the secret. Needless to say, they turned up in due course, tired and hungry, and were severely scolded.

Occasionally while my Uncle John was alive we went to my mother's old home Aberlour House. These visits called for best clothes and best manners, and were of interest mainly for the wonder of the enormous house with its view of the River Spey in front and the huge formal garden behind.

There is no doubt that the three spinster aunts came out tops for their unfailing and generous hospitality to all their relatives — younger married brothers and sisters, innumerable nephews and nieces and, in time, great-nephews and -nieces. We loved staying with them, and they loved to fill their house with several of us at a time. In the winter they moved to a house in Edinburgh. Once, after we had left South Africa, I remember spending Christmas with them, and found it more fun than being at home in Hampshire. Joan Backhouse was also there, and we felt very grown up being allowed to walk the streets of this big city on our own. There was a story that on one occasion a party of my cousins arrived at the station in Edinburgh and could not remember the address of the Aunts' house. However, the taxi driver said 'Oh yes — you want to go to the home of the three Aunties' — such was their fame.

On one of our longer visits to England from South Africa I was sent to a day school in Petersfield for a summer term. The four-mile journey was accomplished in the milk lorry which passed the dark, gaping gateway to Langrish House every morning. It is doubtful whether I learned anything that term. All I remember is the games of cricket and the eurhythmic dancing lessons when the mothers came to watch, and all exclaimed about my brown, sunburnt legs. So once again I was made to feel odd — not quite like the others.

Another unhappy time was when we took a house in Petersfield and Miss Tin came with us. My mother went into hospital for a week or two to have all her teeth out and I was left alone with Miss Tin in an urban environment with no friends. I must have been about seven at the time. It was here that I suffered the only really severe punishment from Miss Tin that I remember. She locked me in my bedroom for a whole day (or at least that's how I remember it). No doubt I had done something very awful — but it seemed unfair and harsh at the time.

The visits to England were always referred to by my parents as 'going home'. This was a common expression to all the colonials in those days, but it made me very angry. My home was in South Africa. I felt passionately about the country and used to write bad verse about it. Why wasn't it *our* home and *their* home?

There was nothing very special about the last trip home from Southampton to Capetown in October 1930, but on arrival things were different. Instead of taking the train straight to Westminster, my father, mother, and I went to stay at Admiralty House in Simonstown, which at that time was a big British naval base. My father had stayed there on several occasions but it was my first visit. It provided me with the opportunity of being taken over a submarine.

Although I had been used to staying in houses with servants and being waited on at meal times, I had never met anything quite as grand as Admiralty House with its large staff including footmen. They haunted and frightened me. After four or five days it became too much for me and I begged my mother to allow me to go home. Being waited on hand and foot was not my idea of fun and there were servants all over the place watching like Big Brother. My father was ill, I knew, but then he had always had these attacks so there was nothing to indicate that this might be the last time I would see him. No doubt under the circumstances my mother was glad to get rid of me and I was put on a train at Capetown for Bloemfontein where Miss Tin was to meet me.

The journey took two nights and a day. Some man was asked to keep an eye on me but he was unobtrusive and no doubt I was unforthcoming. I felt very grand and grown-up travelling on my own at the age of eleven, walking back and forth to the restaurant for meals. South African trains had a very particular smell — not very nice. The black smoke and the smuts were terrible. But I was going home and nothing else mattered. We chugged slowly up the Hex River valley where once or twice in my life I had seen distant snow patches on the hill tops. We crossed the flat low-scrub-spotted Karroo desert and then on across the high veldt with its farms and wide open spaces. At last we came to Bloemfontein with the hill behind it where someone, perhaps in Boer War days, had painted a huge white antelope on a rock high up on the east face. Miss Tin was there to meet me and after changing trains, we were off again for the last sixty-mile run to Westminster, stopping on the way at Thabanchu and Tweespruit with its huge dairy. Finally the familiar beloved *kopje* came in sight. Home at last. The dogs gave me an appropriate greeting and so did the farm manager and his wife, the McKiddies, and my Basuto friends.

Soon after my return, lessons began again. I heard that my parents were on their way but that they had stopped in Bloemfontein. Then on 12 November 1930 I was having an art lesson — sitting at the large table in the nursery with its green baize table cloth, painting a violet. The telephone rang — two long rings because we had a party line — and Miss Tin went to answer. She came back into the room and said nothing until she was standing beside me, and then she told me that my father had died in Bloemfontein. What was there to say? I remember being amazed because her tears dropped on my drawing paper. So I cried a little, too. Then I went out into the garden to try to digest this momentous news. It was frightening, really. How would everyone behave? What was I expected to do?

Within a few days I was taken by train to Bloemfontein, where my father had been buried, and I was then re-united with my mother. Dressed all in black and sitting in a dark, dreary hotel room she was a pathetic figure, and very frightening. In no way could I be a substitute for her loss. It was positively terrifying. My letter to Evelyn after my father's death was almost entirely concerned with what would happen to the farm, the dogs, the cats and my pony, Scot.

The next few months were full of uncertainty. Discussions went on behind my back. Boarding school was mentioned and brochures appeared — an alarming prospect. There was a farm sale and there was a week-end when I was sent to stay with local friends. I enjoyed the week-end, but it was a gloomy summer. Scot was sold for £5 to some people called Helm who promised to look after him well. The McKiddies were left to clear up and to look after the dogs. Not for one minute did I believe that we were leaving for ever, though I took the precaution of bringing some soil from the garden with me. As the buggy took us on the last drive to Westminster station I promised my Basuto friends that I would come back — no tears, just a terrible feeling of emptiness. In July 1931 we made the final Union Castle crossing to England.

CHAPTER THREE

Goodbye to Beauchamp

My mother had decided that our base or home should be in the South of England and not far from the Talbot-Ponsonby relations. House hunting was quite enjoyable, and we finally fell for a newly-built house on a hill just outside the village of Droxford in Hampshire. It was called Pineridge. As far as Evelyn and I were concerned the best reason for choosing it was that we were to own a stretch of the river Meon at the bottom of the hill and the whole 'lot' consisted of seven acres. As the garden had not been laid out, my mother would be able to have fun designing it. The house was not pretty but there was something rather attractive about the idea of its being new, thus enabling us to impose our own personality on it.

My next destination was St George's School, Ascot. The first year was spent in the Junior School, which was in a separate building called The Red House, which had its own charming headmistress who was very kind to me. It was not as bad as I had imagined that it might be, but I had the impression that girls and staff alike thought me rather odd. For one thing I was very well developed for my age, with a large bosom, while my dormitory companions were still scarcely showing signs of physical maturity.

Having had a governess and individual attention I was far ahead of my contemporaries in most subjects — except Latin, which I had never learned before — so the work was not difficult, and by the end of the year I was top of the form without any effort at all. I had been deprived of music all my life and thought I should like to learn the piano, but it was depressing playing scales with one hand when girls junior to me could play quite elaborate pieces. The final decisive factor in giving it up quite early on was that my teacher suffered from terrible body odour.

The following year I was moved with my class to the Senior School where there were about 120 pupils. The rule of the game as far as I was concerned was to work hard and play hard. My cousin Barbara Backhouse was two forms higher than I, but though friendships between older and younger girls were not really encouraged, we were allowed to share a two-person dormitory. My problem was that I had been moved up one form after a term in the Senior School and had left my friends behind. Most of the other girls had come up through the school together. Barbara, on the other hand, had not been there for long.

Between us we managed to cause quite a lot of trouble, and I was forever being summoned to see the headmistress in her study, where she endeavoured unsuccessfully each time to reduce me to tears. If I hadn't disliked her so much I might have behaved better, and she must largely take the blame for my anti-establishment attitude. But Barbe and I had a lot of fun and a lot of laughs. On one occasion I organised a riot in another form room where a scripture lesson was going on, without actually taking part in it. I was so proud of this escapade that I wrote to Evelyn demanding that he come and see me, as I was confined to my dormitory. Many of the girls had seen a photograph of him taken at Harrow, and his solo visit not only caused quite a stir but was very brave of him. He had been instructed to walk up to the front door and I would hang a roll of toilet paper out of the window to direct him to my room. I was in bed with hay fever at the time.

After a while, games as an activity began to pall, especially lacrosse. And here I must admit that it was Barbe who led me astray. We became great shirkers, or skivers, and once hid in the laundry room which was a vast area underneath the main part of the building, filled with huge pipes and boilers. Unfortunately, we were locked in by someone and were faced with the problem of how to get out. There was a lift shaft down to the laundry, but we didn't want to give ourselves away. It seemed a very long time before we managed to attract the attention of a girl passing the lift on the next floor up, who let us out.

The best thing that happened at school was that once a week we had a riding lesson for two hours. The ponies were brought to the front door and about two or three of us went off with Mr Prince, our instructor, to Windsor Great Park. The ponies were about 13.2 or 14.2 hands and were always very beautifully turned out. Tommy 1 and Tommy 2 were look-alikes — black with white blaizes, lively but good mannered. Merrylegs on the other hand was a bright bay and rather hot-blooded. It was my particular delight that I was most often allowed to ride her. She always put in a buck at some stage and I remember only once falling off. She also had the hardest mouth and took pleasure in running away with anyone who couldn't hold her. After several terms my mother asked to have a report on my riding. It came: 'She rides very well, but needs to learn style.'

I took my school certificate when I was fifteen, gaining six credits and a distinction in history, which was my least favourite subject. I stayed on another year in the sixth form, studying mainly maths, the subject I liked best. Then I refused to try for University as I would have had to read maths — which then seemed to me to be a fairly useless prospect. That I did not go to University is the greatest regret of my whole life. It is at University that one makes friends who last a lifetime, and it is also where one learns not just facts but to think for oneself.

After my very solitary childhood in South Africa, where I talked more to the three dogs than to anyone else, it is probably surprising that I coped as well as I did at school. If ever there was an introverted child I must have been one, by force of circumstances. My parents had not been particularly interested in me, the whole family were almost totally undemonstrative, and I had learned not to show my emotions. The main outward manifestation was that I suffered extreme bouts of hay fever from the age of about ten until I was in my late teens when some injections more or less cured me. But all this early repression may have been what caused me to possess a rebellious spirit which still will out from time to time, and which got me into trouble at school.

At Pineridge in the holidays I was painfully shy with grown-ups and contemporaries alike. The kind of parties we went to — such as indoor games at the vicarage — were beyond my range of experience. Evelyn and I would invariably find ourselves partnered off with the wrong sex, since our names are unisex and I was always known as Phil. Even in the hunting field later on I was shy and never spoke unless spoken to, and probably only replied monosyllabically. My mother was also desperately shy, something I only realised later, so this was probably an inheritance.

Among our more amusing activities during the winter school holidays was rigging up our attic at Pineridge as a stage and putting on one-act plays with our cousins, Barbe and Joan Backhouse. These were hilarious affairs, and Joan and I would suffer terrible bouts of giggles. Inevitably Barbe and Evelyn took the lead parts. I was cast as a middle-aged woman, and 'wee Joanie' as she was known, was the maid.

The most positive factor of life during my late teens — and indeed ever since Evelyn's visit to Beauchamp in 1929 — was our relationship.

When we were together the three-year age gap seemed non-existent. We talked of anything and everything. At meal-times, as a mental exercise, we argued endlessly for fun, which infuriated my mother. We dug ponds and made water gardens together at the bottom of the field by the river. We talked for ages while one or the other of us was in the bath — which shocked my mother, but she couldn't stop us. We could read each other's thoughts. When separated we wrote long letters to each other. Evelyn always kept one or two secrets up his sleeve in order to maintain his superior authority by virtue of his age, but we were in fact sometimes taken for twins.

When I left school in the summer of 1935 I was only sixteen. I was still shy and awkward with strangers, but in the family I nevertheless had the bit between my teeth so that when my mother suggested that I should go to some kind of domestic college and learn cooking, the art of running a house, and other useful things of that kind, I was forceful in my refusal. It may have been in my last school report that the headmistress had written under *Behaviour*: 'She carries all before her'. My mother asked what this meant, and I knew that it was not intended as a compliment.

It may have been about this time that my mother found her ideas for my future being swept aside. Evelyn was in Germany — between leaving Harrow and going up to Cambridge — staying with a family by the Starnberger See, near Munich. He wrote long and exciting letters about his activities. The envelopes were almost entirely covered with stamps, leaving only just enough space for my name and address. Why should I not go abroad, too?

So a small school in Neuilly, Paris, was chosen for me, and I was to stay two terms. The school, at the Villa d'Argenson, was run by three sisters, and there were not more than twelve girls of different nationalities. I shared a room with a delightful Swedish girl called Britt Marie Evers, who called me Piliffa, pronounced as the French 'Peeleefa'. She and I conversed in French but, unfortunately, there were other English girls there and, although it was forbidden, we did of course break into English from time to time. I might have learned more fluent conversational French in a family, but I would have missed a wealth of learning in other subjects.

Our lessons in the morning included French language, literature, art and dressmaking. Every afternoon a young woman collected us for some form of educational sightseeing, or sometimes some activity such as skating. We travelled by bus or Métro, and I came to love

Left: Villa d'Argenson, the finishing school in Neuilly where I spent two terms.
Below: Line-up of the girls at the Villa, Easter term, 1936. I am third from left.
Below left: Whisky at Pineridge not long after I bought him.
Left: Pineridge soon after we bought it.

Paris. Going to the opera or to the ballet was the most exciting. Both were entirely new experiences. In those days everyone dressed up for the opera — the women in long dresses and the men in dinner jackets or tail coats. We had no male escorts, alas! But those evenings were glamorous occasions, and seeing Serge Lifar dance for the first time after his return to the Paris Opera House was utterly memorable.

On Sundays we could go to whichever church we chose. During my second term I used to visit the Russian church because I loved the singing and was fascinated by the service.

We had sufficient freedom for my rebellious instincts to be kept at bay. Only once do I remember incurring the wrath of one of the mesdemoiselles. My friend Stella and I were puffing quietly away at cigarettes by the window of her room when the door burst open and the youngest sister put her head round the door. 'Qui fume, qui fume?' she asked, in excited, angry tones.

I was the youngest of the girls at the Villa d'Argenson. Had I been older I might have resented the establishment rules more: but during those two terms I learned a great deal about the things which have interested me throughout my life — such as opera, ballet, theatre and French painting and sculpture.

After leaving Paris aged seventeen, and until the war, I lived at Pineridge and suffered intense boredom — with occasional bouts of social activity when Evelyn was at home and during the hunting season.

While at school I had been out fox-hunting once or twice and had enjoyed it very much, but since leaving South Africa I had not had a pony of my own. At this time my Uncle Ned Talbot-Ponsonby was Master of the East Devon Foxhounds, and lived near Tiverton. I was invited to stay on my own to hunt with him and his terrifying wife, Aunt Dora (always known to Evelyn and me as 'Aunt T-Dora' to differentiate her from my mother's sister). Staying in this large house on my own aged seventeen was very alarming. One dressed for dinner and I remember vividly the long brown taffeta dress with an orange sash which I had made myself and wore every night. After a day's hunting, a mustard bath was prepared for me in case I should be stiff. Driving to the meet we sat in the back of a large car with a rug over our knees and hotwater bottles on our laps. Aunt T-Dora was the aunt who preferred nephews to nieces and had previously at Langrish declared me to be a rude little girl. 'Can't you teach that child better manners?' So it was not surprising that I was nervous.

The banks in Devonshire were formidable. I had only hunted in relatively flat country, with hedges and fences to jump. But the hired horse I rode came from a livery stable in Newton Poppleford and guided me safely over all the hazards, so I had no need to worry about the banks.

It was a year later, on my second visit to these T-P's, that I sighted Whisky for the first time. A lean, wiry horse, 15 hands, with high withers and good strong bone, bright chestnut with a crooked white blaize and two white socks. He seemed ideal, my uncle's recommendation was not be questioned, so my mother bought him for me.

He arrived at the station at Droxford at the end of the hunting season and was put out to grass in the field next to our house. I rode him occasionally during that summer, and soon he filled out and became quite fat. I thought that I was going to be able to look after him myself, but when I had him in the stable in the autumn for a short while and he started to get fit and lively, it became apparent that I knew nothing about stable management, so we engaged a young man as a groom/gardener. But it was not long before I realised that he was no better at managing Whisky than I was. This high-spirited little chestnut horse could be quite frightening.

Later we heard that Whisky had been so unmanageable and wild as the mount of one of the whips in an Irish hunt that he had been shipped to England where, in a livery stable, he was hunted three times a week and worked so hard that he hadn't the energy left to behave badly. We also discovered that he was only five when I first had him.

The young groom was soon replaced by an older man called Stubbington, or 'Stubbs', who had worked at the Hambledon Hunt kennels. He exercised Whisky when I was unable to do so, and I knew that Whisky would not be beaten about the head or maltreated if he behaved badly. He may have been more than a handful for me at first, but in spite of my being bucked off several times — especially on the long rolling track across the golf course on Corhampton Down — we soon got used to each other. I had three marvellous seasons, hunting him twice a week with the Hambledon hounds, riding sometimes as much as ten miles to a meet. He was the fastest horse in the hunt and would jump anything I put him to, often giving a lead to others. His dam was by the same sire as Hyperion who won the 1933 Derby, so he was running true to form. His only bad habit was that after hunting he would jog all the way home. He was very highly strung, like most chestnuts, and I could never cure him of this jogging. The only really bad experience I had with him — and it was not his fault — was when he put his foot in a rabbit hole at full gallop. He turned a somersault and I was thrown on to my head — which resulted in concussion. The people who came to my rescue were surprised to hear me repeatedly saying 'I want Whisky.'

I don't remember seeing a kill while out hunting. It was the chase that I enjoyed. Galloping across country on such a special little horse was exhilarating and exciting. There was the art of knowing every field and every gate and short cut in order to keep up with hounds. An elderly neighbour had given me some local maps with a scale of 4 inches to the mile which helped greatly and were interesting to study. A good day's hunting gave me the sense of freedom that I had not experienced since South Africa.

When Whisky's spring coat came in after his winter clipping it shone like burnished copper. His walk was springy, fast and smooth, and there was a breathtaking exhilaration in riding such an alert creature and feeling one with him. He always carried his head high, and his closely pricked ears nearly met in the middle. He was above all else my pride and joy, and he won high praise from many who saw him out hunting. On a windy day on the downs as he pranced on the turf, I thought of the quotation:

> Proudly he trod, his golden mane up-flung
> Like a carved frieze against the morning
> His mane wind-frayed, his nostrils blown
> Wide in a heedless ecstasy.
>
> *(Margaret Stanley Wrench)*

These were indeed moments to be treasured, as also were the rides through the hazel copses in spring, among the carpets of primroses, anemones and bluebells. Such moments were to become even more treasured when war broke out in 1939.

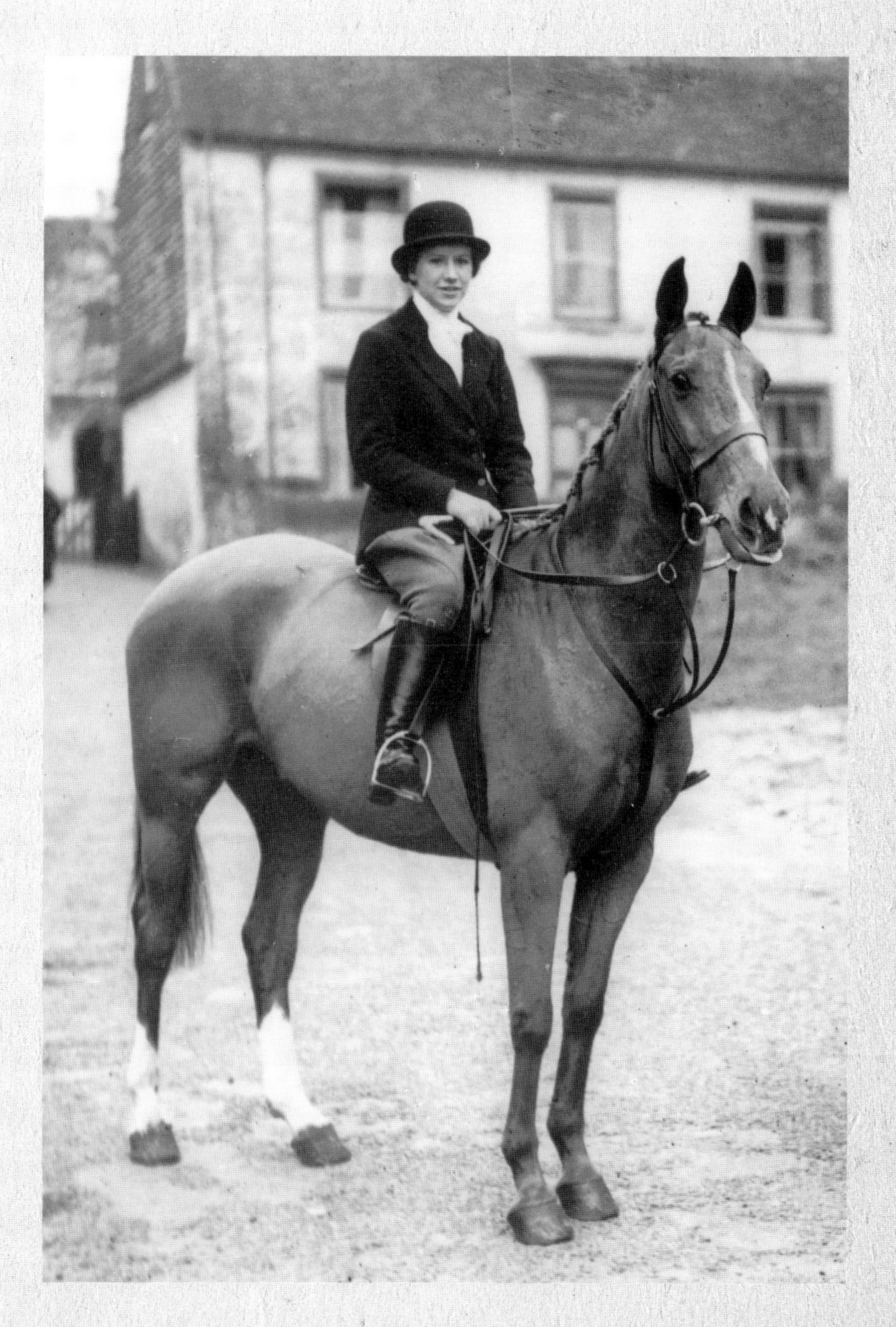

Whisky in his prime at a Hambledon Hunt meet in Hambledon Village.

The Cambridge Clique

For me the years between leaving school and the beginning of the war have a curious quality. There were some wonderful occasions, like the holidays abroad with Evelyn's Cambridge friends, the visits to Cambridge when Evelyn was up at Trinity reading Modern Languages, and the May Week Balls. There was the time I was invited to a performance in German of the *Urfaust* with Evelyn as the student, Antony Part as a very sinister Mephistopheles and Kit Dodds as Faust. The crocuses were out on the backs. There was a Masked Ball. There were idle hours in punts on the Cam. I was very young and it was all very romantic. While writing this I sense the nostalgia, and can conjure up the scent of Pheasant's Eye Narcissus, which for some reason I associate with Cambridge days. But the May Week Balls were perhaps the highlights. I shocked myself by buying the model dress on the stand in Dickins & Jones for the wildly extravagant sum of five pounds. I hardly dared tell my mother. She only approved my visits to Cambridge when she knew that I would be lodging in a convent which provided bed and breakfast, or staying with friends who lived in Selwyn Gardens. It was clever of Evelyn to arrange these visits so well.

Evelyn owned an ancient Austin Seven which he bought for about seven pounds and which was nicknamed Jerry. On one occasion it died on us at Rickmansworth on our way to Cambridge. We took a bus to Watford where we prised an old radiator from an Austin van on a car dump, and took it back to the Watford garage where Jerry was to be hybridised and picked up on the way back. Then came the problem of hiring a self-drive car to take us on. Hire car garages didn't like the look of Evelyn. He was a typical undergraduate with baggy grey flannel trousers, and was therefore considered unreliable. We tried several places without success, then Evelyn said: 'Well, you go in on your own and try'. With some trepidation I tried another garage. Though younger (I think I was nineteen at the time), at least I didn't have the undergraduate stamp, so eventually we hired a Riley sports car and drove in style and at speed to Cambridge where we

Performance by the Cambridge German Society of the Urfaust.

Above: Kit Dodds as Faust.
Left: Antony Part as Mephisto.

were due for lunch with Evelyn's ex-tutor.

Fringe benefits from my brother's time at University meant meeting friends of his who became my friends. There was 'the clique', a delightful group mostly reading modern languages. They were a romantic lot, forever quoting German poetry and with an enthusiasm for life which was infectious. Inevitably I fell in love with a marvellous man. They were all special people but this one was more special. Later, after the war had started, I panicked — realising suddenly that I didn't want to be tied for life. The agony of telling him remains with me, because I loved him. However, we remained good friends.

At that time the dentist in London loomed large in my life. A terrifying man, very bald with rimless glasses, who said 'Ummm' in a long, slow murmur as he probed and drilled deeper, and who prophesied that I would have false teeth before I was forty (he was wrong). Later, he passed me on to his more friendly junior assistant whom I continued to consult until I was over fifty. The good thing was that my teeth were so bad that frequent visits to London were imperative. These weekly visits, which somehow tended to coincide with University vacations, were a wonderful excuse to see friends in London without having to explain everything to my mother. She was kind enough to go on paying my dentist's bills until I finally left home after the war. It was also a good excuse for a day off during the war when I met friends and 'did' a show.

One of our London friends had a sister who was 'doing the season', and her mother, breaking all the debutante rules, was kind enough to invite me and some of the Cambridge girls to her Ball in their house. The girls were all supposed to be 'Debs' of the year. In those days there were dance cards with numbered dances and tiny pencils attached. Men approached their chosen partners ahead of time and sometimes asked a girl for several dances. Names were exchanged and pencilled in. This was fun because you could play your card close to your chest. In the early stages there was an element of risk, because by holding out hopefully for your favourite partner you could be left with a few gaps and consequently the dullest man in the room, or being a wallflower when the time came. The Debs' mothers sat around and watched and gossiped. Our hostess, I recall, was not pleased with her son because he booked several dances with me. It was his duty she told him afterward to have danced with as many different girls as possible. All our special Cambridge friends were there, including Evelyn. The occasion was especially memorable because it gave a great boost to my self-confidence. Surprisingly, I felt that I had been a success. I was not just 'Evelyn's little sister' and I was left with a warm glow of happiness for being a person in my own right.

As soon as I was seventeen I started to have driving lessons from the local blacksmith, Mr Alfred Taylor, and took my driving test in Southampton in the summer during a bout of severe hay fever. I was afraid that the examiner would think I was crying. However, I passed first go, and almost immediately had to drive to London in my mother's Morris to collect her after some minor operation in St George's Hospital. The blacksmith's warnings remain with me yet: 'Beware of the kiddies and the cows!'

My mother generously allowed me to drive her car, and was even very tolerant when we had a minor accident in a narrow local lane, grazing the side of an on-coming car. But she was very strict about my boyfriends. Reluctantly she allowed me to go to Glyndebourne to hear *The Marriage of Figaro* with one of 'the clique' who collected me from Droxford. We were to stay overnight at a small hotel in Seaford. I read the libretto beforehand, and in the afternoon we walked on the downs and saw the Seven Sisters White Cliffs. The performance was magic and we came home the next day. It was all very romantic but quite above board. When I admitted to my mother that I had been unable to persuade my friend to let me pay for my room — that he had in fact paid for both rooms — she was furious. I had never seen her so upset. She wanted to refund the money. I said that he would be hurt, and we argued furiously. I can't remember the outcome, but I think (and hope) I won. Later this friend gave me for my birthday a complete recording of the whole opera with the same company, and I came to know it very well. I have seen it only once since, but would love to hear it again at Glyndebourne.

It was rather an aimless existence, living at home with the occasional tennis party in the summer or dance in the winter. There was the water garden down by the river where Evelyn and I spent many hours together when he was at home. I was supposed to look after it during University term time. We always had a dog — first a Cairn terrier called Brownie — rather large for her breed and with only one eye (she had lost the other having distemper as a puppy). She had belonged to my naval relations, the Backhouses, emulating Nelson with her one eye. She was a dog of great character and we adored her, so that when the Backhouses took a house near by in Fareham they allowed us to keep her. To our shame, however, Evelyn and I once went to their house to play tennis and forgot to bring her back. My mother was upset, fearing that we had lost her for good, and we had to collect her the next day. Rather extraordinarily, while driving back to Droxford sometime later my mother found a dead Cairn terrier with only one eye lying in the road. Very distressed, she brought the corpse home only to find that it was Brownie's double. Brownie was at home safe and well.

Later I had a golden cocker spaniel bitch called Penny, who was rather beautiful but quite stupid and lacked the character of the old Cairn. Having dogs meant walks. And we did walk. My mother and I, or Evelyn and I, or friends and I, plus dog, walked the downs for miles. My favourite landmark was Beacon Hill with its stand of beech trees (now gone). Evelyn's favourite was Old Winchester Hill, now a National Nature Reserve.

So the time was taken up with these various activities and I read a great deal — the whole of Shakespeare, *War and Peace* again, other long novels and some biographies. Whisky, my horse, was still the most important person in my life. But where would such a life lead? I didn't know many people locally and, as the Cambridge friends achieved their degrees and took up jobs, I became more isolated, and more frustrated in a quiet sort of way.

Left: Brownie, the Cairn terrier whom we took over from the Backhouses.
Below: Evelyn and Antony Part on top of a mountain in Austria.

Holidays Abroad

Real highlights of the pre-war period were holidays abroad, which were always in the summer. Not nearly as many English people went to the Continent in those days. Mass air travel did not exist. Our first trip was when I was sixteen, to Achensee in Austria. On that occasion my mother came with us. The party (Cambridge connections again) consisted of two families and three more young people. Ten in all.

During this trip in 1935 the Salzburg Festival brought me my first real-live concert. Bruno Walter was conducting. Perhaps suitably for one of my age and experience, it was 'light music'. The two particular pieces which I remember and which thrilled me were Schubert's *Unfinished Symphony* and Strauss's *Tales from the Vienna Woods.* I watched Bruno Walter lift just one little finger to bring in the special sound that he wanted, and I marvelled. This was the beginning of a lifelong interest in music, and Evelyn was to help me with the records of Beethoven and Brahms (and Wagner, who was not among my favourites) which he brought home. The outdoor production of *Everyman* at Salzburg was equally momentous.

At Achensee we scrambled up all the surrounding hills: nothing very high, but enough to give me a taste for scrambling. As well as Achensee and Salzburg, we visited Innsbruck and Lofer. Around Lofer the hills were higher, with rocky peaks. Here on one of our climbs we spent a night in a mountain hut (the Erfurter Hütte). We slept on a *matratzenlager* — all seven of us on one mattress, with Evelyn and me providing the dividing line between the sexes. I disgraced myself by talking in my sleep and suddenly sitting bolt upright in the middle of the night and announcing dramatically 'HUT'. After this there was much teasing, but next morning I was one of the smaller party who climbed the neighbouring peak. My mother was not so amused by the *matratzenlager.* She thought it was highly improper. She did not enjoy the holiday much — she was the odd woman out. So when another holiday was planned for the following year, she agreed to allow me to join the same group again for a further Austrian adventure.

We travelled by train with no more luggage than we could easily carry, and stayed at small inns and climbed higher hills than the year before. Our first base was Tristachersee, then on to Maria Luggau on the Austro-Italian border. It was more or less the same party as the previous year with a few additions, but no Evelyn, who was bicycling to Budapest with a friend from Harrow. Also in our group were an Austrian Baron and Baroness who, though middle-aged, contrived to liven up the party considerably. The Baron was particularly fond of practical jokes, and we retaliated by sewing up the bottom of his nightshirt and making an apple pie bed. It was sad to learn later that the Baron was a Nazi who took an active part in the Anschluss of 1937, when Hitler annexed Austria.

A letter from Nora, the Baroness, dated 20 April 1938 is worth recording:

> Baden bei Wien 20th April, 1938.
>
> 'My dear Phil — I have not forgotten, but really the last months and the time before were so exciting and also trying — really I couldn't thank you sooner for your letter. Ninety per cent are as glad as we are that the horrible nightmare before the 10th March [the Anschluss] has passed — not only that our hopes, we secretly nourished for years have been fulfilled, but also we stood before at war in our country, just like in Spain; only imagine that. Now already work is here for our people and money for the quite poor ones; I'm sure in a year you see no beggar. Perhaps you are not interested in this, as I know England was never for the idea; but if you make a clear logical resumé you will see that the country and people which were against didn't help us either; only this great and clean man, (Hitler), with his fanatic love for the whole United Germany could and would do it. And my dear Phil how are you? My husband has much to do and also our baby car works for the idea. . . .'

She goes on about a dog that they have adopted.

Those holidays were very happy and carefree, with lots of climbing, walking and dancing to local bands. We climbed Monte Peralba (2693m) which was just over the Italian border, and nearly got into trouble with frontier guards who spoke no English, but who in the end agreed to escort us to the top. Speaking no Italian, we could at least use musical terms if we wanted to go faster or slower. For the more energetic climbs there were usually only about four of us. Once, Antony Part and I were caught in a thunderstorm high up in a circle of stone peaks set around a beautiful small lake with the thunder reverberating and the lightning streaking down to the rocks. It was among these lovely Austrian rocky peaks that my passion for mountains developed.

That year for some reason another girl and I came home to England on our own — a complicated train journey which involved several changes. It gave us time to spend part of a day in Munich, where we watched the sunset from the bridge over the River Isar and drank beer in a genuine *bierkeller*.

By 1937, some of the previous year's party were doing other things, and I was invited to join the Part family on a trip which was to begin in Jugoslavia, continue to Budapest, and end in Vienna. The group consisted of Antony Part, his sister Rosemary, his mother (who for some reason was always called 'Aunt Part') and a delightful friend of hers known as 'Uncle William'. It wasn't easy to persuade my mother to finance this trip, but she did. The opportunity for such a 'grand tour' of Europe was too good to miss, and the way we were to do it was far from expensive. Our train journey took us through Germany (Aachen is always remembered from each of those years for the very welcome cold beer), Austria, and finally into Jugoslavia, crossing the border near Klagenfurt, staying a few nights in Dovje just across the border, and then going on to Kranjska Gora in the Julian Alps. There was a fairly large hotel and a few other small houses as well as the railway station. From here Antony and I climbed some local peaks before deciding that we must attempt the highest mountain in Jugoslavia, Triglav (2863m). We made our plans carefully, consulting local people and acquiring a map which showed the footpaths.

In the morning we took the train one station along the line to Dovje, carrying our food and extra clothing. Then after lunch we set off on a long, hot walk up a beautiful tree-clad valley where there was a clear stream and a high waterfall. We were aiming to reach the Staneĉeva Koĉa by nightfall in order to make the final ascent of the peak next morning. By 4 pm we had reached a hut at the end of the valley, and from there we started our ascent: at first a very steep climb among trees, and then two hours' solid rock-climbing. There was a tricky bit of ice to negotiate, which delayed us, so that by 7 pm we were still a long way from the next hut and the sun was setting. A glorious sunset, but no time to enjoy it.

It was quite exciting tramping through the huge basin below the peak of Triglav in the half light across several patches of snow and a very badly marked path with never a sight of the hut. It was also quite frightening, especially as we knew that the hut was a small one, or *Koča*. When we finally reached it in the dusk we felt unwelcome. There was not the usual Austrian greeting of 'Grüss Gott'. The people were dark, swarthy, tough and sinister-looking. They were not really hostile: merely surprised to see us arriving so late.

After some goat's milk and some soup we went to bed. At 3.40 am we were awake, and by 4.40 we were off again. Soon afterwards the sun rose, lighting up the peak of Triglav. Then as it shot up over the shoulder of another peak, its rays turned the ice all around us to pink and then to gold. We reached the top at 6.30 am and stayed there until 7, marvelling at the magnificent view. We were so cold that we put on all the clothes that we were carrying, including our pyjamas.

The memory of that expedition is one of sheer happiness — among the 'greats' of my mountaineering exploits.

From Kranjska Gora we went first to Ljubljana, where we spent one night, then by early morning bus to Sušak (or Fiume) where we boarded a boat called the 'Triglav' which was to take us to Crikvenica. From there we were to travel, again by bus, to Plitvice. But it was Friday the 13th. The boat left Sušak late, and there were no buses out of Crikvenica that day. As the two older members of our party had decided that this part of the journey would be too strenuous, and had gone direct by train to Plitvice, there were just the three of us, Antony, Rosemary and I, stranded in Crikvenica until 6.30 the following morning. We found a room for the night, and next day we had an extremely long, hot and tiring drive in a rickety old bus. However, Plitvice was certainly worth all the effort: a splendid spectacle with its seventeen blue lakes all joined by waterfalls. There were early morning walks before breakfast and there were huge meals (described in detail in letters to Evelyn) at amazingly low cost, which must somehow have restored the energy I was using up, because I never put on weight.

Left: On the steps of the hotel at Achensee. Back *(l to r) Evelyn, Antony, Sheila Macleod, the Baron.* Front *(l to r) Sybil Corbett-Winder, me, Rosemary Part.*
Below: Setting off for the Erfurter Hütte.
Right: In Jugoslavia. Antony, Rosemary, Uncle William and Aunt Part.
Far right: Convalescing at Carezza, 1938.
Below right: My first attempt at back-lighting: Sybil Corbett-Winder outlined against the mountains.

Another boat journey from Crikvenica took us to the beautiful island of Rab, but it was merely a break in the voyage to Split and I regretted so much that we were not staying there. There was much sightseeing in Split — including Diocletian's Palace and the huge bronze statue of Archbishop Gregor of Ninsk — as well as some hilarious bathing from the popular beach, which was not very attractive.

On a tiny, weekly, and therefore very crowded, steamer we went on down the coast to the small fishing village of Podgora, which I am sure must by now be a famous resort. In those days there were just three small hotels, miles of olive groves, and lots of heavily-laden donkeys driven by fat men wearing fezes.

Just as we were arriving at Podgora, after a magnificent sunset, a sudden gale of terrific force got up. The groups of people waiting on the little pier took to their heels and fled, and the pier light was smashed to smithereens. After half-an-hour, all was quiet again and we were able to land. Our hotel was right by the shore, and we ran down to the warm Adriatic Sea and swam and rested and swam again by moonlight. It was an idyllic place, and we had a whole week there before going on to Dubrovnik. Although we were only in Dubrovnik for one day and two nights, I recorded the fact that this city and Rab were the two places in Jugoslavia which *must* be re-visited. There were a few English tourists — about the first we had seen since leaving home. Apart from the fascinations of the city itself, this brief visit was memorable for my first taste of oysters.

On we went with our 'grand tour' — a day in Sarajevo, a night in Belgrade, and then happily to Budapest. How lucky to have seen it before World War II! Long letters record a list of the places we visited, together with details of the *pensions*, the cafés, and the swimming pool with its artificial waves (the *Gellert Bad*).

The train journey on to Vienna was not uneventful, as we travelled in the same coach as the Duke and Duchess of Windsor, and at one point got a wet shower through the window when they chucked some water out.

I was the only person interested in horses so I went alone to see a rehearsal of the Spanish Riding School in that marvellous building, the Hofburg. At this stage of our travels I think we were all getting a little tired, and I remember nothing about the journey home, though

I think we stopped once more on our way back through Austria.

The summer of 1938 brought the Cambridge climbing enthusiasts together again, with various girl friends and a few others, for a holiday in the Dolomites. This time there were no 'adults'. My mother decided to stay with her sisters in Scotland, and Evelyn was once again with us. We stayed at Carezza al Lago, and I shared a room with a delightful girl called Petronilla. I had a cold at the beginning of the holiday which lingered on as a cough and, finally, after a good soaking in the rain on one of our climbs, became so bad that poor Petronilla was woken at night by me sitting up in bed coughing and gasping in terrible spasms. Evelyn arranged for the hotel to call a doctor. He provided medicines, I started to lose weight, the cough worsened, and I became thinner and thinner. The holiday was nearing its end, and then one day the doctor announced that I had tuberculosis and was unfit to travel. As my father had died of TB, it was all rather dismal, but I lay out on a deckchair covered in rugs, with the beautiful wild autumn crocuses on the grass all around, and kind friends came and sat at my feet and talked to me.

A cable to my mother reached her just as she was stepping into a car at Aberlour in Scotland, bound for John O'Groats. She immediately came South to collect her passport, and was with us at Carezza al Lago the day before the rest of the party left for England. It was decided that my mother and I should move to a hotel in Merano, where most of the guests were undergoing a grape cure (slimming diet): five kilos of black grapes a day, and nothing else for two weeks. A German doctor visited me at frequent intervals, giving me a course of injections. I lay on the balcony eating grapes, looked longingly at the hills, practised my rather bad German, wrote long letters to Evelyn, and wondered if I should ever be able to climb mountains again. Even to walk a little uphill made me catch my breath and cough in frightening spasms.

The long letters which I wrote to Evelyn are quite amusing to read, even now. Here is a quote from one dated 11 September 1938, from Merano:

> 'We have just been watching a pilgrimage which took 20 minutes to pass the hotel gate. It began with masses of men in pairs with lovely green braces and Austrian hats, with priests among them. Then came some small boys carrying candles and then a shrine carried by men with bright blue stockings and lederhosen. Behind these a row of small girls in their Sunday dirndls and wreaths in their hair, also carrying candles; behind these, women in black with pale green, blue or yellow aprons and shawls (lovely), small girls in white, then the Bishop or whatever he was in gorgeous robes under a canopy carried by more priests, and after that a band (Alpine) followed by an endless stream of people all in their Sunday best. When the band wasn't playing they were all chanting and telling their beads, and once when there was a traffic jam the small boy swinging the incense didn't notice the halt and hit one of the priests such a blow in the back! They also had two enormous banners which were a bit difficult to manipulate because of the electric wires which cross the road overhead just by the gate. It was all a lovely sight, but I doubted if they would ever get into the cathedral'.

My mother and I visited the nearby Schloss Tirol which had some splendid ancient marble sculptures in the doorways. The castle was the home of the famous 'Ugly Duchess' (Margaretta von Maultasch) whose portrait was painted by many artists including Leonardo da Vinci. By this time the rumours of war with Germany were becoming ominously persistent, and before long there arrived a telegram from the Scottish Aunts: ADVISE YOU RETURN HOME IMMEDIATELY. They cabled extra money for First Class tickets, and we took the train home, passing through Munich just at the time when Neville Chamberlain was actually there making his peace proposals to Hitler.

At Munich we had to change trains and as we sat in the station restaurant wondering if the Germans would be hostile to us, a waiter came up with a vase of flowers and said: 'Roses for the English ladies'. The other tables did not have flowers. We were touched by this gesture and of course Chamberlain arrived back in England waving his piece of paper with its now notorious 'peace in our time' message.

And so I returned home to my beloved Whisky, of whom I had been thinking a great deal. He needed to be brought in from grass and exercised ready for the hunting season and I thought I was the only person who knew how to get over the problem of girthing him up when his skin was still soft and tender after his long holiday in our field. He hated his girth being done up, and had thrown me and our first groom several times during the two previous seasons.

First of all, however, there was the question of my health. My mother insisted that the advice must be the best. For doctors and dentists she was always prepared to pay what was necessary. The King's physician in London looked at my X-rays, examined me and — contrary to the original diagnosis — said I had whooping cough and bronchitis. The after-effects of the whooping cough would take time to disappear but I could certainly go climbing again next year if I wished. And so I went out riding again with a cough that echoed round the South Downs, but happy to know it was not TB.

A Special Mountain

By the summer of 1939 many of my climbing friends were working or were married. I still had no idea what I wanted to do with my life, and was suffering the frustrations of living at home in the country. So I made a plan with Petronilla Maitland, to go to Switzerland. We both had the same feeling for mountains, we had shared a room in the hotel at Carezza the previous year, had laughed a lot, and generally got on well together.

The Bernese Oberland seemed to offer the best scope for our virtually non-existent talents, and we booked in at the Hotel Bellevue, Kleine Scheidegg, at the end of July. Enquiries at the hotel revealed that guides were available from Grindelwald and we arranged to meet one, Fritz Steuri, to discuss plans. He was a wiry, good-looking, quite small man of about thirty-six. Our enthusiasm for climbing amused him. 'Why don't you climb Tschuggen?' he asked. Tschuggen was a small, rocky peak on the other side of the hotel away from the great ice walls of the Jungfrau and its attendant giants, the Mönch and the Eiger. 'Tschuggen!' we cried with furious disdain. 'We've been up there'. So he agreed that if the weather was all right he would come back the next day and try us out on the face of the Jungfrau.

Our trial climb, roped up, on the snow and ice of the Jungfrau didn't get us anywhere much — except higher in Fritz's esteem which was what really mattered. The weather drove us back before we had climbed very far, but Fritz promised to take us up the Mönch (13,468ft).

It was an exhilarating climb — even though, rather shamefully, we took the mountain railway up as far as the Jungfraujoch. My letter to Evelyn describes it thus:

> 'We started off [from the Jungfraujoch] towards the Mönch in soft snow and bright sun, with this lovely view southwards over the snow, and big woolly clouds in the distance. After three quarters of an hour or so we roped and started abruptly upwards behind two parties of Germans. It was not by any

means a difficult climb, but very thrilling for the likes of us. We went up to a ridge and then followed it the whole way up. At first it was very deep new snow on a rocky path, *very* steep, with every now and then some rocks to be climbed up and over. The Germans were slow and bad on the rocks so we passed both those parties. Then there was a bit of snow ridge which we climbed just on one side — about a foot away from the edge. The slopes on either side fell sheer away to the bottom with rocks sticking out half way so that one couldn't really see the bottom, but as yet we weren't very high. But soon after we proceeded up a long, long bit, fairly steep at first then getting steeper, where we were absolutely on the knife edge of the ridge. A mist was coming up by then, but one could see the slopes on either side falling away terribly steeply for a long, long way. Another lot of rock, then we were once more plodding up the ridge. By this time it was so steep we had to go half-time ordinary slow mountaineering pace. Step cutting was tiring work but we had caught up two more parties of men. The steps were quite large — I mean apart — and it was an effort for a small person like me. The loose snow was swept in our faces by gusts of wind. It was then it suddenly occurred to me how awful it would be coming down, and for several moments I wondered if I should ever manage it, although I was not actually in the slightest frightened going up — only cautious, trying to plant my feet firmly and not lean forward. At the top of the ridge there was a small snow peak and there we passed the other two parties and went on along a cornice. A voice behind me, that of a guide of an English party, said "This is overhanging here!" I laughed, I couldn't help it. We were about 3 feet from the edge of the curl-over which was leaning away from us and the slope fell away steeply on the other side, but it wasn't far from the top and we were first on the top in the virgin snow! The view appeared in bits as the mist swept up and over the shoulder of the Mönch, but it was a terrific view when it did appear. The Jungfrau gleaming in the sun with a vast range of mountains behind her, the Eiger a little below us but very impressive, and such cloud effects as I have never seen before. Fritz shook us each by the hand and congratulated us in a most touching way. You can have no idea how wonderful it was — or

Right: Petronilla and me on the summit of the Mönch.

Left: Fritz Steuri, our Swiss guide, on the summit of the Wetterhorn.
Above right: The Wetterhorn from the path down to Grindlewald.
Below right: View from the summit of the Wetterhorn. By now I was beginning to take photography quite seriously.

> perhaps you can — but then you have never been told you could never climb again. It was such a terrific view and it had been such a good climb'.

The descent was evidently much less frightening than I had anticipated and on arrival back at the Jungfraujoch we explored the curious hotel which is built into the side of the mountain, and the ice palace behind it, which was full of ice carvings:

'One little room had a piano, hot water pipes (!), stools, a cask of beer and a table all carved from the ice with nice round pillars to the roof.'

We had planned to climb the Jungfrau the following day so we spent the night in bunks and were to be called at 3.30 am. But at that hour it was still snowing and we had to give up the idea. Fritz said that the Jungfrau was really just a trudge anyway and we would do better to plan for climbing the Wetterhorn when the weather improved.

The weather did improve at last, and Fritz was right. The Wetterhorn turned out to be an even more memorable climb than the Mönch. We walked down the winding path to Grindelwald, where we were to meet Fritz. The sun shone in a cloudless sky, the Alpenrosen were in bloom, and people we met greeted us with the time-honoured 'Grüss Gott'. After Kleine Scheidegg, which is about 5,000ft above sea level, Grindelwald was hot and airless.

The worst part of the day was walking through the town and along roads and paths to the bottom of the Ober Grindelwald glacier. Once there, we began to go up, and finally crossing the glacier it was onward and upward steeply and lengthily to the Gleckstein Hütte where we arrived soon after 4 o'clock streaming with sweat and panting with thirst. Fritz prescribed tea, and in no time at all he produced a huge pot like a watering can from which we slaked our thirst. Swiss huts provided masses of pairs of boots which you put on when you took your own off. They were vast things with rope soles and lined with felt. They were so large for us that we looked ridiculous, and we laughed and laughed. Everything was perfect. The clumps of two different kinds of gentians were a brighter blue than we had ever seen. There was a carpet of other flowers; the sunset was spectacular. The hut looked out over the glacier on to the Schreckhorn and the Kleine Schreckhorn with their exquisite snowy peaks.

That night there were fourteen guides with their respective parties — which was apparently a record, and was due, no doubt, to the fine weather. Even Fritz, normally so pessimistic, said that the weather was 'all right'. We slept in a two-bunk room because Fritz was doing the same and the hut owner said the mattress places were crowded and would be uncomfortable. Fritz teased us and wouldn't tell us what time we would be getting up. First he said 2.30, then he said 3 o'clock, then 1.30, and his final words were: 'Maybe 8 o'clock, maybe not until 10'. We slept like logs and were awakened at 1 am. We couldn't believe it — but the hut was buzzing with the sound of voices.

By 1.45 we were off, the fourth of several parties. This was real magic. The stars were shining and there was moonlight on the snow peaks of the Schreckhorns. We started off on a steep, rocky path, Fritz in front swinging the lantern. There were three other lanterns visible on the mountain above, and behind us, when we stopped to rope up, we could see the light of the hut. From rocky path on to a crusty, smooth, snow-covered glacier, on to rocks again, another glacier to cross, steep this time; and sometimes one sank through the crust, or someone in front had done so with one leg and there was a deep hole. On to snow-covered rocks on a vertical ridge where at 4 am we left the lantern under a rock and passed the other three parties. Up the ridge, over snow and rocks — and only one star left in the sky. I don't know why, but the guides joked about us all the time. Halfway up we left the ridge and crossed a gulley, which was tricky work. There was ice on the rocks and as it was not really light one could only tell by the fact that the ice-covered ones were a bit darker. It was, perhaps, the most difficult bit. After more rock we got into a *couloir* where it was hard work cutting steps. The other parties went on ahead after this to give Fritz a rest from step cutting. The ice or hard snow came rushing down on us from the people ahead as they cut the steps, and we ducked to miss it. It was a long haul up that *couloir*, but at 6.15 we reached the saddle where we got into the sun and saw over the other side. There had hardly been time to watch the sunrise when the snow peaks behind us glowed orange and the rocks turned red. But now we were in the sun and here we had a rest and some breakfast.

Fritz set off for the summit with us in the lead again. I was next on the rope with Petronilla bringing up the rear. The rhythmic sound of step-cutting, the ice tinkling down the slope and the occasional bits of snow in my face if I got too close were exciting and the summit peak was very steep. At last we were on the snow face of the top itself, almost as steep as a wall, before the final overhanging cornice like the crest of a wave. We broke through the crest and reached the top at 7.45 am. The view was magnificent. The sun shone, there were photographs, the guides joked and Fritz was happy — especially so because he had got 'his girls' to the top first.

We only had a quarter of an hour on the summit, because it was important to have crossed the *couloir* before the heat of the sun set off any avalanches from the rocks above.

I led on the first steep slope — so steep that one could hardly prevent one's behind coming in contact with the snow even if one did lean forward. I had the problem, too, that the steps were far apart for me. The best part of the descent towards the hut was the glissade on the snow-covered glacier. Back at the hut at 11.15 we again drank quantities of tea, had an early lunch and then set off to look for edelweiss. We found it growing in a deep gorge near the glacier, so we had to rope up again. Then came yet another thrill — a marmot! It was the first one I had seen, and I was amazed by its loud whistle.

We were too tired to do the three hour walk back to Scheidegg from Grindelwald so we took the train. My letter to Evelyn says: 'It was awful saying goodbye to Fritz. He was very strong and silent, and we were silent, too, because we were so tired. His favourite climb is the Schreckhorn so I must come back and do it next year'. I had big ideas. The plan for 1940 was to climb the Schreckhorn, the Finsteraar, the Lauteraar and the Fiescherhorn, not to mention the Berglistock (because it had such a nice name) — and Evelyn would come too.

The War and Cows

So much for my climbing plans! Not long after my return from Switzerland, war was declared. As we heard the news on our old Pye wireless my mother gnashed her teeth in anger and cracked her dental plate, which upset her even more.

My immediate reaction was to go and join the Land Army. Life at Pineridge was dull, the war was a challenge and the women's services were not for me — I had had enough of uniforms at boarding school. So two days later I started work at a farm just down the road, for Mr Sylvester, who agreed to pay me 16 shillings a week for milking the cows twice a day. He usually forgot to pay me, and after a week or two I had to go and ask for my wages. I was rather nervous about doing so but his sour disbelief that I was making the grade and sticking it out made me bolder as the months went by and I knew I was earning every penny.

The hours were from 5 to 8.30 am and 2 to 5 pm varying slightly according to the number of cows we had in and whether one of the three milkers was off sick. And of course it was a seven-day week. It was about ten minutes, walk down the hill through the village of Droxford to the farmyard which was next to the Manor House. The head cowman was one of the nicest people I have known. He was very strict and we were very formal — 'Mr Hounsham' and 'Miss Talbot-Ponsonby' we remained — but he had a tremendous sense of humour and pulled my leg unmercifully. The third member of our team was a boring and boorish old man who was rather deaf and dreadfully slow.

Mr Hounsham fetched the cows in every morning and we all met by lantern light in the farmyard and chained up the cows in their stalls as they came in. There were anything from thirty to forty cows of a mixed herd being milked at one time, and it was all hand milking. We each had a lantern and started in separate sheds, as these were spread around the midden. Before milking we washed the muck off the udders and back legs with warm water which the cowman had heated in a copper (a huge bowl like an inverted bell). He lit the fire under the copper before fetching the cows. After each cow had been milked she was let loose on to the midden in the centre of the yard.

At first if a cow put her foot in the milk bucket, or kicked it over, I used hopefully to sloosh the offending white liquid down the gutter drain — but I could not get away with that trick because the milk appeared some minutes later, shining white, in the manure in the middle of the yard and Mr Hounsham never failed to notice and to chide me with an amused grin. Sometimes I was able to pull the cow's leg out of the bucket without upsetting it, but the straw and dirt which appeared on the strainer gave me away. One was not supposed to be careless enough to allow the cow to do such things. When cows kick they kick forward, not back like a horse and it was possible to form the habit of keeping one's left arm up against the hind leg to forestall a kick.

As I improved at the job I became quite fond of some of the cows. They varied enormously in character and ease of milking. There was a big black and white Friesian who gave her milk easily and in great quantities. Here was an opportunity for thought or day dreaming. 'It occurred to me under the black cow . . .' became a regular preamble to any ideas I produced at home. Another favourite was a little crossbred Guernsey/Shorthorn. I was heartbroken when a friend came to buy her to help out with the milk rationing, but delighted when she was returned two days later because she simply would not give her milk to a stranger.

A cow would be milked for anything between nine months and a year then, being dry, she would just stay with the herd or perhaps in another field until she calved again. The calves nearly all went to market. On coming back to the farmyard to be milked, each cow remembered the stall she had been in before, and if that one was now occupied by another there would be a lot of pushing and shoving until they were sorted out and had their noses in the cow cake at different stalls.

Once, I arose an hour early by mistake and finding the place deserted went into the bull's stall which had a deep litter with a bigger heap of straw in the corner. Here I lay down and slept until Mr Hounsham arrived. This episode amused him enormously and he delighted in telling the story, often slightly embroidered for dramatic effect.

He was not at all amused one Christmas morning when I forgot to take the plug out of the milk-cooler at the right moment — and so much milk ran away that the whole midden looked as though there had been a snowstorm.

Only once was I chased by a bull. It was a young one which was running with the cows. I was assured that at that age he was quite safe. But one day after milking I was collecting the herd from a small paddock beyond the main yard in order to drive them down the road to the field. I had a small stick but nothing else. The cows streamed out quietly through the gate into the yard but the bull refused to go and started to walk round me in ever decreasing circles with lowered head. Mr Hounsham was in the dairy, and the old man was slowly shovelling manure in the yard without looking up. Just in time, Mr Hounsham looked out of the dairy, took in the situation, seized a pitchfork and came tearing across the yard to rescue me. After that the bull was kept in.

Often I guiltily secreted a handful of cow cake nuts in the copious pockets of my khaki Land Army overcoat to take home for Whisky. Years after the war when Mr Hounsham came to visit me at Slimbridge I was very touched when he told me that he knew all about it and had turned a blind eye. Obtaining oats, bran, or even hay, for a hunter during the war was very difficult. Luckily we were able to make hay in our 5-acre field and to keep Whisky in a small paddock for part of the summer.

My one and only experience of a Magistrates' Court was in connection with the cows. I had been walking ahead of the herd as we took them up the winding road through Droxford. The old man was driving them from behind. It was my job to slow down the on-coming traffic and to keep the cows spread out in a long line so that cars could get past. One morning a car failed to slow down, and it hit a small black cow. She collapsed in the road with her back legs broken, unable to move. It was very horrible and the driver of the car blamed us. The cow had to be destroyed. Mr Sylvester asked me to give evidence, and some weeks later I tore down to the court on my bicycle only to be sent back because my head wasn't covered. The deaf cowman had to have all the questions shouted at him and quoted distances in chains which caused the members of the Bench to look quizzically at each other and Mr Hounsham and me to giggle quietly.

At first I resented wearing Land Army uniform. Our Commandant was a friend of my mother's who organised not only the Land Army but many other services in the area. She was a lovely person with a family of eight children, the youngest one being adopted. She

Left: In uniform, 1940: me in Land Army dungarees, Evelyn as a Signalman, RNVR. Later he gave me his bell-bottoms (see picture on front cover).
Below: Cartoon sketch by Belinda Panton, my fellow milkmaid, in 1943.

WOMEN'S LAND ARMY
HAMPSHIRE COUNTY COMMITTEE

Certificate of Merit

awarded to

Miss F. P. Talbot-Ponsonby

for PROFICIENCY IN MILKING
in the
County Milking Competition, 1941

Class B

F. G. Bullock
County Dairy Instructress.

Date August 1st 1941

J 3530

Right: Proof of my war effort.
Below: Resting at mid-day on the ride to Polperro.

didn't care what I wore, but when clothes rationing came in I was glad to have the green jersey, the gum boots, the thick three-quarter-length khaki overcoat and even the hob-nailed shoes. I could never come to terms with the baggy riding breeches, but wore my own. Jeans were unknown in those days.

The hob-nailed shoes got me into a spot of trouble. I used to run down the steep road through the village at 5 o'clock in the morning with my nails ringing on the tarmac. It came to my ears that the villagers were talking. They said that I woke them up in the morning, that I was not pulling my weight in the war effort, that I had not 'joined up'. I was furious. Five until 8.30 in the morning, two until 5 in the afternoon seven days a week and then there was still the housework and the vegetable garden — all that seemed to me to be quite hard work.

At one stage I contracted a particularly nasty rash and skin infection on my hands and arms and was given a temporary respite from milking. A local friend who was, I think, a part-time VAD kept horses, and we decided to take a riding holiday down to Polperro in Cornwall where she knew people who ran a riding school. We decided that her brother's big bay army horse was in better condition than Whisky, and she rode her lovely grey hunter. We planned the ride carefully, spending two nights at pre-arranged places on the way and on the way back. We had a wonderful week hacking around Exmoor, which was new country for me. On the way back her horse went lame so we had to be collected from Salisbury, which was sad, but the holiday had been good.

There were occasional days off in London to meet the Cambridge friends — and even occasional visits to the dentist. Those days had a special quality, and were to be treasured, because we never knew when — or even if — we would see our friends again. In the early years of the war people were dispersed around the country, but not necessarily abroad. There were still theatres, concerts, occasional dinners and even dancing (on a smaller floor) at the Hungaria restaurant.

It was during one of those visits to London that I first met Peter Scott. A friend of mine, Kit Dodds, took me to a dinner party at the basement flat of a friend of his called Dosia. It was a memorable night for two reasons: the other one being that it was my first experience of a London peasoup fog. Bad enough by day but far worse at night in the war-time black out.

The door of the basement opened. A girl said: 'My deaf uncle is here. I'm sorry, but you will have to shout at him'. He looked quite young and handsome but we duly shouted in his ear. Not for long. The two girls present collapsed in fits of laughter. The 'deaf uncle' was Peter Scott and one of the girls was the beautiful Jane Howard to whom he was then engaged to be married. After dinner we played a silly game. A name was picked with a pin in the telephone directory. The person who could keep up a telephone conversation with the stranger on the other end for the longest time was to be the winner. Amid such talent I was quite terrified, but luckily for me the party broke up before my turn came round.

Sometimes I stayed with the Backhouses in their large house in Sloane Gardens. One of my cousins had warned me that if there was a noisy raid I was to be careful how I behaved. My Uncle, who was First Sea Lord at that time, was rather frightening and apparently he had been very cross with one of his children who suddenly got under the dining room table during a noisy raid. Later in the war, at the time of the V1s (doodle-bugs) and V2s, I discovered that the draught-excluder at the bottom of the front door made a noise exactly like a doodle-bug in the dark.

During the early part of the war Evelyn was living at 99 Baker Street, where he joined the Fire Service on 1 September before war actually broke out. But in January 1940 British Petroleum for whom he had previously been working re-called him to their evacuated offices at Walton-on-Thames. He was trying, at first unsuccessfully, to get into the Navy, and then the Marines. Finally he was accepted for the Navy — for the good reason that his father, his grandfather and his great-grandfather had all been in the Senior Service. He began as an Ordinary Signalman and looked very handsome in his bell-bottomed trousers and jaunty sailor's hat.

After a spell at sea on HMS *Dunedin* he was entered for a commission, which he obtained on 1 January 1942, coming top of his class. Shortly after he left the *Dunedin* she was sunk off West Africa by U124. When Evelyn had joined the ship a friend had lent him an eighteenth-century caul. Cauls are reputed to bring good luck to seamen, and the ship's company were thrilled, but eventually he had to return it to its owner. Not long afterwards the ship was sunk — sheer coincidence of course.

In the meantime because of his excellent German Evelyn was selected for Naval Intelligence — little knowing that he would not go back to sea for the duration. He reported for work at Bletchley Park on Friday 13 January, and thereafter suffered considerable frustration in his attempts to return to sea again.

Between 1939 and 1942 he was often at home for weekends, when not at sea, and life was always brightened by his presence.

At the time when children were being evacuated from London my mother agreed to take in three of them. A group duly arrived at Droxford: and my mother collected three brothers from the village hall. Their mother seemed glad to be rid of them. They were aged five, seven and eight. We had endless trouble with the middle boy who used to climb out of one side of a three-panelled bedroom window in order to get in at the second one along — a highly dangerous occupation. Then there was the little one who didn't like his food and, if made to eat it, immediately threw up on the dining room table. Rationing was tight and food was scarce which made it seem worse. (At breakfast I used to be so hungry that I filled myself up with lettuce.) Their mother visited them occasionally for the day, bringing expensive toys when what they really needed was new shoes or socks or other clothes. My mother was noble and caring, but I was less tolerant.

The days were long and physically demanding and I was forever hungry — more so as rationing grew stricter; two ounces of butter don't go far in a week. At one point during harvest time I succeeded in getting an extra ounce of cheese a week. I wasn't actually harvesting but I did plant potatoes on the farm — aided by, and in charge of, a gang of thirteen children who had been let off school for the job. We all had buckets which had to be filled from bags which were further and further away as we proceeded down the furrow. It was a case of dropping one in every two feet — literally two-foot lengths. One, two, plop; one, two, plop; and soon I discovered that the children were so tired that their potatoes were going in one every two yards because it meant fewer trips back to the bags.

Ironically, the evacuees returned to London before the bombing began in 1940. The war had been 'phoney' for so long that eventually parents wanted their children back and the government agreed to allow them to return.

We had an interesting time with air raids in and around Droxford.

First during daytime in the late summer of 1940 we saw the dog fights high up in the blue sky. Then in 1941 came the bombs. Portsmouth and Southampton were not far away and were fairly well defended with anti-aircraft guns. The enemy planes which came over us had been forced inland by the coastal guns and were dropping their bombs indiscriminately before returning home.

Letters to Evelyn at that period make exciting reading. Incendiary bombs at night were frequent, and when the area lit up like daylight we felt very vulnerable. The worst incident was in 1941 when a shower of incendiaries fell — some in the garden, one through the roof of the house (it miraculously ended in the water cistern), many all over the drive, and two close to my precious hay rick. I rushed down and extinguished them, on the way throwing myself face down by the dung heap, which was also on fire, as I heard the swoosh of another lot coming.

'Packets' of incendiary bombs were called 'bread baskets' and you could hear the swoosh as they passed near by. Land mines, too, fell unpleasantly close, and there were many nights when the house rocked and the doors and windows rattled with the blast. We had a rather inadequate home-made dug-out shelter in the garden, which could hold about six people. I have found only one record of its being used — when various members of the household including friends and neighbours and my mother's cook gathered there, clad in all sorts of strange attire.

The only good thing about being so physically exhausted all the time was that I was able to sleep better through all the noise than most people. One of my letters reports a lone raider which flew low under cloud over our field, machine-gunning as it went and dropping its bombs not far from the railway station opposite our house. For us the worst manifestation of this disaster was that it was too much for our cook who, to my mother's great regret, gave in her notice. An elderly widow who lived alone across the road spent many nights in our house, sleeping or resting uneasily under the stairs while my mother prowled around or went down to the village first-aid centre known as 'the point'.

After our evacuees left we had two other people billeted on us for a while. The husband was working, but the wife was with us all day. They experienced one of our worst incendiary raids and helped in the extinguishing game.

One night in June 1941 an elderly couple, great friends of ours who lived just across the valley near the railway station, suffered a direct hit on their house. Mercifully they escaped through the back door, but the house and garden were a write-off. The next morning my mother immediately went over and brought them to Pineridge, where they stayed for some time. Their daughter, a special friend of Evelyn's, was a wartime nurse and was later married from our house.

As though I did not have enough to do I bought a New Forest pony foal from some gypsies who were ill-treating her and her mother. Friends bought the mare, and for £5 the foal was mine. I called her Mitzi, and almost from the start, she turned out to be a little devil. In my innocence I thought this could be cured by kindness: but as the months and years went by, rougher methods had to be adopted for breaking her in. Her worst habit was biting. She would approach with teeth bared, so that even catching her was a problem. I enlisted the help of a friend in the village, Barbara Wade, who looked after several horses for neighbours, and between us we finally broke her in. She grew to 12 hands. I bought a pony trap and a set of harness, and we trotted round the lanes — but not for long, as I had too many other things to do and she was not at all a loveable animal. I sold her to a strawberry grower for £25 before leaving the Land Army in 1943.

There was also Penny, the golden cocker spaniel who had succeeded Brownie the Cairn. As Penny was mine, she slept in my room. Her coat was much the same colour as Whisky's. Her ears were so long that they met at the end of her nose and slopped into her dinner bowl. She had furry feathers between her toes and not a brain in her narrow head. She never learned road sense, so I could only occasionally take her with me when I was riding. She disgraced herself by producing black and tan illegitimate puppies, one of which we kept for a while and called Pengo after the Hungarian coin. I think we gave her to Stubbs (our groom/gardener). Sadly her mother, Penny, developed a tumour at a fairly early age and had to be put to sleep.

Extracts from Letters to Evelyn

19th August 1940

'We have had some exciting raids this week. So many that I can't remember which happened when. During one I had just taken the cows to the field and was near Clark's [the grocer's] watching with a lot of the village. It was very cloudy but a low plane suddenly appeared almost overhead and everyone said "Spitfire", as I believed, but afterwards discovered that this was the one that had machine-gunned Sylvester's men working on a rick, and various other people as well. Another clear day I saw 15 Germans coming over this way amid clouds of anti-aircraft shells bursting, and one turned tail and disappeared seawards with smoke pouring out of it. Then, the day we had Mrs Panton and Belinda to tea there was a terrific battle. I think they were horrified at us standing on the stoep watching because whenever the siren goes Major Panton marshalls his family into the shelter while he stands at the top and delivers a running commentary. Belinda actually was thrilled to the core by the sight of so many German planes — mere silver specks roaring tremendously, miles high — and especially when we saw some of our fighters pursuing two planes out towards Hambledon. The Corhampton gun (I think) has been in action. The place to see things from is Corhampton Down I gather. From there people have seen planes crashing, balloons dropping like nine pins, and bombs exploding. Bishops Waltham had a bomb in the brick works. I will tell you all the stories of damage caused around when I see you. You may never get here because the trains are awful. Henry Bruce took five hours to get from Aldershot to Alton on Saturday . . .'

Later same letter:

'I have just been down at the pond weeding the garden. I was much engrossed in that immense task when the rattle of machine gun fire made me jump nearly into the pond! It sounded as though it was just over Merrington's [the garage on the road next to our field]. Actually the plane must have been somewhere above the Kennels [the Hunt Kennels half a mile away]. I took cover under the ash tree! One of our fighters appeared about 10 minutes later and flew round beneath the blanket of cloud but the Germans had gone off above the clouds . . .'

This was the famous long hot summer of 1940 and the letter goes on:

> 'The field is practically bare and Whisky comes and watches me weeding, looking over the fence all the time waiting for tit bits. I offered him my weed bucket and he and Beryl [a mare we took in for grazing] munched dry columbine and plantains. It's pathetic! I cut the grass on the far side of the ponds for them the other day. I thought I might as well blunt the hook before you came back!'.

September 1940

> '. . . I have got to go to a Land Army tea on Sunday 22nd to receive my second good service badge from the WLA (Women's Land Army) county officer at Mrs Bruce's. According to Molly [a neighbour who was staying with us as there was an unexploded bomb in her garden] I was instructed to go in my "knickers". [A reference to the baggy uniform breeches.] Tough lad you must be to do semaphore after inoculation. Why did the officer hit you in the back??! . . .
>
> 'Did you know there was an attempted invasion last weekend??!!'.

December 1940

> 'Jerry celebrated last night in much the same manner as last Saturday when you were here. Though the house didn't rattle so much, there was a bigger, redder glow over May Hill and there was a terrific fire over that way to be seen at 5.30 this morning. You know Stubbs had a mother and child, evacuees from London? Well, she said it was worse at Soberton than in London. So she left by the early train this Sunday morning. Silly fool! Apparently she lives near the docks in London, anyway. "Ever such a nice person," said Stubbs, but she has not returned the £1 he lent her to go home with, and left all her bills unpaid. She had complete jitters during the raid and clung to Stubbs! He couldn't get her out from under the table to go to bed. It must have been a funny sight! [Mr Stubbington was a very shy, quiet person.] The nearest bomb — the one during supper — fell about 600 yds from Stubbs' house and according

to the inspector who had Stubbs with him when he found the crater, it was the biggest we have had around here — a 1,000-pounder.'

20th April 1941

'The Blitz came to Droxford last week on the night of the 17th. I know bomb stories bore you, but after all it is something to write home about and parts were quite funny. I was just getting into bed when the siren went and the noises began. I heard the Singletons go down shortly after, and as the noise was so terrific and sleep seemed impossible at that stage I joined the party on the stoep. It was a grand fireworks display, and it was mostly nearer than Pompey. Flares were dropping by the half doz., baskets of incendiaries swooshed and lit up over Denmead way and on Teglease, guns boomed and the planes sounded low and close. The poor house rattled like a lunatic. At 11 o'clock it calmed down a bit so we all retired to bed, but it was a brief lull and it was soon raging in full force again. I have never known the house heave and shake like it did all through the night. I was dead tired having had a strenuous day — milking 5–8 am, potato planting near the Workhouse from 9–12 am, milking 2–4 pm and weeding the Brita path from 5–7 pm. Once I was just dozing off dreaming I was planting potatoes (one, two steps, drop!) in the gravel path when there was a terrific bang. I never slept properly and at 2.45 am I was lying cursing those bloody jerries, when there was an almighty rushing, whistling, swooshing sound, a plop and innumerable small bangs. I instinctively covered my head with the bed clothes (!!), Penny for the first time whimpered and leapt on to my bed, I jumped out of bed then and looked out of the window for a second to see the whole field, garden and everything lit up with fairy lights! Grabbing my coat and slippers, there came again that awful swooshing, so with a feeling of "this is really the end", I doubled up beside the bed as the bombs exploded and everything rocked. However, we had just been missed, the bombs were just the other side of the road from the top of the drive. We all dashed downstairs and opened the front door. It was as bright as daylight and there were incendiaries everywhere, in among the daffodils, one beside the garage wall, in front of the kitchen window, on the tennis lawn, in both gardens, dozens in the field and in the football field, in Mrs Dean's, in Major Hulbert's wood — everywhere! Mr Singleton seized one of Mother's small sand bags and threw it on one just by the drive, it exploded violently and I later discovered that he burnt his hand. I did likewise on the one by the garage but luckily it had already exploded. Mr S and Greta, I believe, then tried what Mr S will call the stomach [stirrup] pump but found it not much use, so Mother, he and I armed ourselves with the three long-handled shovels and tore round wildly putting out the fires. They are the devil to get out even if only on grass or earth. Mr S kept saying "they are sure to make us the target now we are all lit up." Down in the village we could see a great red fire blazing from somewhere, a house we thought, but actually a big straw rick in Sylvester's yard. The fire engine had gone to Denmead so it continued to blaze and burnt right out. Mitzi was in a state of terror and Mr S nearly let her out, she knocked down a post and grazed her face and leg. I went down the vegetable garden putting out the things as I went. I was just struggling with the manure heap which had caught fire, when again I heard "swoooosh". With a promptness which impressed me vastly, I flung myself down in the dewy grass and waited. It was another bread basket just above the Kennels, and all the time there were planes zooming horribly low overhead. Three times I had to go flat, once on the gravel which was most uncomfortable! Another lot of incendiaries over beyond the SW's (the outsiders fell in their gardens) and some bombs by the station, one direct hit on the station cottages by the beech walk. Our luck was amazing, because the incendiaries were literally all round the house, and though one went through the potting shed roof it didn't explode. Wilsmer had one in his shop and they had to break in through the window to put it out. There was also one through the Mill Cottage roof. There were two inside the fence round our hay rick which I put out before they spread. While down there I tried to see if Whisky was alright but by that time the fires were burning out and though his field was full of them it was dark and I couldn't see or hear anything. I called, but not a sound. Needless to say the first thing I did when daylight came was to rush down there, but he was flourishing, though jumpy, and had knocked a bit of the fence down.

'After we had put all the fires out, Mr S insisted we should go down to the shelter as the rick in the village was still a blazing beacon and the noise was still going on, so Greta [our house maid, who not long afterwards left to join up], Mrs S, Roberty [our cook], Penny and I all packed in there while Mr S sat on the steps and Mother went down to the point. I did laugh down there! At 4.15 there was a lull and Mr S put the kettle on and we all emerged and drank tea in the kitchen — and so to bed, but not to sleep because the guns started again and Penny immediately got nerves and insisted on coming into my bed where her tummy rumbled so much that sleep was impossible! The All Clear went just before 5, and as I went down to milk there was an endless stream of cars going back to Portsmouth — people who sleep in their cars out of town. There are four craters just above the workhouse on the left of the road — just missed the cows I suppose — two time bombs opposite us, one beside the Farnols' new house. They are of course evacuated, and Farnol moaned to me that they wouldn't even allow him to retrieve his teeth! The other is just about opposite our drive, but they don't look like going off. There was a land mine at Hill Place, and lots of craters round about the station here.

'It was all very exciting! But I think once was enough. Drinking tea in the kitchen, I laughed at the party all in their strange attire (Mother had gone down to the point) Mrs S in a pale pink hair-net, Greta in a very all-enveloping brown one, Roberty with a white pigtail hanging down her back and no teeth, and all with coats over night garments. I had put on my trousers before going to the shelter. Greta remarked that she had often wondered who would look the best on such an occasion! Mr S has been very sorry for himself ever since, with his burnt hand and bad blisters on his heel. It has certainly given the village something to talk about as you may imagine. I don't think I had been so tired since after a May Week Ball! I had to go and plant potatoes again next morning too, but only went for two hours instead of three. What's more, Hounsham had to go to a funeral that afternoon, so there was double work to do with the cows, and it was nearly 5 by the time I

> got home to tea having milked fifteen cows while Farnol milked twelve. There were six school boys of about 10 and 11 to help with the potato planting and I was put in charge of them. Mercifully they behaved quite well! But I was amused, because they had been released from school to do it and they started work with tremendous enthusiasm, but it is tiring work carrying buckets full of spuds and walking in the soft earth, and before long they were beginning to put them in about 2 yards apart instead of 2 feet!'.

During 1942 occasional lone raiders dropped bombs in the vicinity, and we were all constantly preoccupied with the news which came through on our old Pye radio. Evelyn was safely installed at Bletchley, which was one less worry: though strangely I had always had great faith that he would survive his period at sea. The war in North Africa was intensifying, and the battle of the North Atlantic was raging. The U-boats were taking a heavy toll and the Allies were suffering very heavy losses at sea. There was pressure for a second front in Europe. The best news that year was the victory at Alamein, in October. On the home front things went on much as usual.

It was at this time that Evelyn gave me the chore of casting in plaster a head of a young girl which he had modelled in clay while on one of his leaves. Before the war he had been to evening sculpture classes at the London Polytechnic and I had helped him to cast some of his works. The problem with this one was that the little girl had tight, curly hair. Picking the plaster out of the curls was a terrible job and I wrote complaining letters to Evelyn about all the other things I might have been doing, like riding Whisky. But the object was achieved, and later I saw the finished result sitting under the stairs in a vicarage near Martin Mere.

An exhibition of Evelyn's sculpture at Pineridge, entrance six-pence, with a notice on the front gate brought in a number of visitors, and a small sum was raised for the war effort. The exhibition included a stone polar bear and several heads cast in plaster and painted — but not, at that time, of Hilary.

In 1943 two important things happened to change my life — Evelyn was married and I left the Land Army.

Evelyn had once said that he had no intention of marrying anyone from Bletchley. He was at that time modelling in clay the head of Hilary Curtis who worked in his department, and not long afterwards he announced his engagement to her. I had met her on a visit to Evelyn when he was billeted in Buckingham, and we were all very happy. It was to be my fourth time as a bridesmaid. ('Three times a bridesmaid, never a bride'.)

Rising at 4.45 am became an increasing strain. I set two alarm clocks and then reached a stage where I would dream that they had gone off, and I would leap out of bed and get dressed only to find it was 3 am. By 9 at night I was so tired that I could not keep awake and by 1943 my social life had dwindled to a minimum. The previous year my refusal (through tiredness) of an invitation to a ball may well have lost me a husband and probably was a major factor in determining my whole future. At the time when I refused I did not know that it was a case of 'Either . . . or . . .' Just as well perhaps.

The rash on my hands and arms from milking was a recurring problem, which became quite serious. Even the cowman was getting me down. A change of scene was definitely necessary. Evelyn suggested that I should go to Bletchley Park. A friend who lived in Hambledon was also working there in the WRNS and she, too, urged me to go. But the question was how? I still did not want to join the services — and that was not the way in which to be accepted at Bletchley. Would Hilary mind? Hilary actually helped. She had a friend working in another department who was prepared to pull a few strings. I gave in my notice to Mr Sylvester, and not long afterwards I was called to London for an interview. This was my first introduction to top level security. I was staggered — but so impressed by the secrecy that I scarcely remember anything about the interview except that I came home with my head reeling and my lips sealed.

My mother gracefully assured me she could manage on her own. Whisky could remain in our field, and Barbara Wade would keep an eye on him; anyway, I would come home for some weekends.

The period between jobs was memorably pleasant. Air raids were becoming few and far between; the Allies were preparing for the invasion of Europe, and there was a lot of military activity on our main road. The station at Droxford was said to be of some significance: it was in fact where, later on, Churchill was to meet General Smuts, General Eisenhower and other war leaders.

I went for long quiet rides on the downs, and time was of no importance. 'You may never have so much freedom again', my mother said.

From Bovine to Clandestine

On 15 June 1943 I left home to begin working at Bletchley. My personal life was to be centred round Buckingham, where Evelyn and Hilary lived in a tiny semi-detached thatched cottage, lying at right angles to the Brackley road out of Buckingham, with a small garden in front and two outside loos and a coal shed at the back. The other half — two up and two down, unlike theirs, which was only one up and one down — was occupied by Hilary's friend Daphne, Joan who had introduced me to my job, and another girl. I stayed there for a short time while one of them was on leave, then moved into a billetwith a postman and his wife in a council house on the other side of the town. Though they were very kind to me, the evenings after supper were rather lonely. My landlady told Evelyn: 'Miss Ponsonby only looks really happy when she has lit her cigarette after supper'.

The bus for Bletchley Park was scheduled to arrive at the end of the road at 8 am. For about a month I worked from 9 till 5, getting back to my billet about 6 o'clock; the buses stopped to drop or pick up people all round the countryside, so the 12-mile drive took about an hour.

During this period I managed to visit London several times to meet friends and go to the theatre or a film. The station at Bletchley was conveniently close to the Park, and transport for the various shift workers was often available at certain times of the day. At that stage the job was very dull, and recreation was definitely needed. As a new girl, I was the coffee-maker and the filing clerk, but I also learned to use a typewriter.

My colleagues were all female, with the sole exception of the General who ran our particular department. Poor man, it must have been hard to be the only male in a hut full of women. He was small, serious minded, very quiet, and absolutely charming. His second in command was a formidable lady, Miss Montgomery, who had known my father when he worked in Naval Intelligence in World War I. She was concerned with the organisation of the shifts, and could be quite frightening. Once, I asked for some days off because I wasn't well: the pressure of work around D-Day having reduced me to a quivering mess. She glared at me and said: 'I suppose you aren't shamming'. I was horrified. How *could* she think I would do such a thing? A friend of mine has reminded me that she wore clean white paper cuffs on her dress every day, and I think of her always in grey. But she was basically kind and wrote a friendly letter to me after I was married saying that my father would have been proud of me.

After about a month I was promoted to shift work. One week of 8 am to 4 pm followed by one week of 4 to midnight meant that the changeover from evening to day shift was very short — off at midnight and back at work at 8 am the following day.

And that included a drive of half an hour or more each way. The long changeover was better, as one came off at 4 pm on a Sunday and did not start until 4 pm on Monday. Days off could be saved up and arranged two at a time which could even occasionally mean a weekend if added to a long changeover.

By October one of Joan's companions had left and I was offered one of the three bedrooms in the cottage next to Evelyn and Hilary. The only running water was in the sink in the living room, and we cooked on an ancient coal-fired black kitchen range. Having a bath was quite a ceremony: the tin bath was set up in the living room and filled from a large black kettle. Since we were often on different shifts it was not too difficult to have the kitchen, or indeed the house, to oneself.

Hilary was expecting her first baby — nicknamed 'Humble' before he was born. Babies did not appeal to me very much, apart from the interest of having a nephew. Andrew arrived on 26 February 1944, weighing only five and half pounds and not without problems. My diary simply records 'Humble born'. However, it was not long before the family were all back in the cottage, and there came the terrifying day when Hilary had to go to London to see her dentist and I was left to be nursemaid to this tiny creature. He must have been a very good baby because I remember very little about it except the anxiety beforehand. This was only the beginning of many times when I was called upon to look after some of my nephews and nieces — but until I had my own it was the only time I was left in charge of a very small bottle-fed baby. When I think about it, I realise how trusting Hilary was.

My diary records remarkably little for 1943 and 1944. There were days at home when cousins or friends came to tea, but my real life in

Left: The Thatched Cottage, Buckingham.
Below: Evelyn with Hilary and red-haired Christopher.
Right: Winifred peering over the stable door of our luxury accommodation at Adstock, 1944.
Below right: Sketch of Little Adstock where we lodged before moving to the stables near by.

Little Adstock.
Adstock, Bucks.
Albert H Knight
12/38

Buckinghamshire remains blank except for the different shifts marked in pencil.

Eventually a complete change came about in our living arrangements when Joan went abroad and the other girl left. I was then the senior partner or 'landlady'. The first person to join me was Anne Watson, a very dear friend who lived near Droxford and whom I had known for some time. We then needed a third, and Evelyn suggested a girl called Winifred (she later changed her name to Catherine), who worked in his department. Not long afterwards Evelyn and Hilary moved to a slightly larger and more comfortable house in Padbury, not far from Buckingham. Once the cottage next door was empty, the owner decided that he wanted to take possession again, and we were forced to move.

After searching the neighbourhood, Winifred and I found some wooden loose-box stables in Adstock, between Padbury and Bletchley, which were being converted into living accommodation. We liked the village — it was conveniently close to Padbury — and we were able to preserve our independence. At least, we thought we were: but it was to be weeks — even months — before we could move in to the stables. In the meantime we were to be accommodated in 'The Big House'. Sadly, we left the cottage on 24 June 1944 and installed ourselves in the beautiful old house known as Little Adstock, a low, thatched, half-timbered building which stood in the centre of the village.

Once again we were really in a billet — though a pretty extraordinary one. We each had a large and comfortable bedroom. We had meals in the dining room. And there was a delightful elderly French widow called Mrs Newton who cooked delicious meals for us. I am greatly indebted to her for teaching me how to make an omelette. Before the war I had almost no experience of cooking, so I was really learning the hard way. Though Hilary had taught me how to make a treacle tart, the food in the cottage had been very basic.

Our landlord, Captain Archie Pelham Burn, was an interesting and eccentric man. He was ex-army, with a very active mind bent on practical application. The conversion of the stables into living accommodation occupied a lot of his attention, but he also liked making changes at Little Adstock. Our relationship with him was always very formal. We called him 'Captain Pelham Burn' (or 'P.B.' behind his back). I was 'your Ponsonby friend' to Winifred and she was 'Miss Illingworth'.

Winifred and I were more frequently than not on different shifts. Sometimes one of us would return from work to find that the dining room had moved from one part of the house to another and it was difficult to discover where it was. Even more disorientating and disconcerting was discovering that one's bedroom had been changed and all one's personal belongings moved from one room to another. All these strange happenings meant that we were forever leaving notes for each other in an effort to get things straightened out. We never knew what might happen next.

Archie Pelham Burn sometimes had visitors from London. Perhaps the most interesting was the actor/manager Campbell Mitchell — who was rather larger than life and very colourful. He had to be introduced to the other inmates of the house even if they were already in bed and half asleep — not an experience which I enjoyed. Later 'P.B.' altered the house again so that he could have part of it to himself. 'I am more or less living at one end of the house and having my meals separately as I want to be more independent and be able to ask my friends in' he wrote in a note to me after I had been on leave for a short while. Mrs Newton was very hurt and upset, but she still fed us. One of Winifred's notes labelled 'POISON' records the fact that she had been given a bad egg for supper, so perhaps the quality of our meals deteriorated.

At last, on 10 August, we were able to move into the stables — a long creosoted building lying at right angles to the road at the end of the village, with the cemetery just beyond the far end. Originally it had probably been four loose boxes. Now it consisted of a living room, two bedrooms and a bathroom. A concrete 'apron' ran along the front. The rooms still had their double stable doors (top and bottom opening).

Our greatest friend in Adstock was a splendid Irish character — called Paddy of course — who came to our rescue as a handy man when things went wrong, which happened all too frequently. He was very deaf, which made things difficult, but he had a wonderful sense of humour.

My worst memory of the stables was going to bed after a day shift when it had been raining incessantly. Winifred was on evening shift, and was expected back some time after midnight. Along the concrete in front of the stables there were huge black slugs and we had discovered that the best way to deal with them was to sprinkle salt on them. Also, by stepping carefully they could be avoided.

On the concrete floor in the bedroom was a rug, and on another rug by my bed lay my slippers; the bathroom floor was concrete, and as the roof leaked in a few places, warm slippers were necessary in cold weather.

I put a foot into my slipper. Inside was a large, cold, slimy, black slug. So I screamed. I have always hated slugs. It was a real nightmare. For some time I sat on the bed with my feet off the floor, shaking with horror. Then I padded back along the wet concrete once more to fetch the salt tin. I searched the room and sprinkled a defence line of salt all round it. A note to Winifred on her bed asked her to come and see me on her return so that I could report the danger. She was amused, though sympathetic, and later we laughed a lot about the incident. From then on salt became an expensive item on our household budget as the whole building was ringed with salt inside.

We must have reported our concern to our landlord because a note from him says: 'Paddy is hard at work filling up the holes where the enemy enters. I share your and Winifred's dislike of slugs; but after having been in the East they take a back place in comparison to snakes.'

Our days in the stables were incredibly happy ones. Even the villagers wanted to know why we laughed so much. There was never a dull moment. We had mice in the kitchen. We set a trap, and when a mouse approached while we were eating a meal we could not bear to see it caught, so we frightened it away.

We had a competition to see who could make the lightest 'five-minute pudding' — a baked sponge which we ate with golden syrup. Winifred's rose the highest, but was deemed unacceptable because she had overdone the baking powder and it tasted revolting.

Central heating was rare in most country homes in those days, but the cold in a house is nothing to the cold in wooden stables which have no ceilings. My bed was covered with innumerable blankets and an eiderdown, but it was still cold. Our landlord kindly lent me a fur rug made from jackal skins (known as a *karos* in South Africa) which was heavy but certainly helped. There were mornings when my frozen breath hung in icicles off the fur above my face.

At the end of November, Winifred went off to Edinburgh where her mother lived, to have her appendix removed. She was away several weeks, so I had various friends from the office to stay from time to time. Life at Adstock was seldom uneventful. Winifred's first letter to me from hospital begins: 'Well really I don't know what things is coming to I'm sure — no sooner have I turned my back than the most extraordinary things begin to happen, all I can say is no cistern never went pouring out boiling water and steam when I had a say in the stable! . . . I think you were terribly brave to tackle the Sadia (water heater) single handed — I should probably have leapt into the bath and covered my head with the new curtains or something . . .'

Her next letter begins: 'I am so sorry to hear that the roof blew off while you were sleeping, I hope it didn't disturb you too much — it's annoying too about the side walls falling like that — but still, you will still be able to get a bit of shelter from the two end ones, and the damaged chimney won't smoke the room out so badly now that there is a little air in it (of course I know that your letter telling me about these latest disasters hasn't actually arrived yet; perhaps you have forgotten to post it? — but I know it can only be a matter of time) . . . I adore getting your chronicles of catastrophe — it would almost be insulting to call them anything so ordinary as letters'.

Certainly the Sadia did blow up. It was frightening. And something did happen to the roof — because I remember Paddy hammering it when I had been on night duty and was trying to sleep during the day.

Winifred's absence lasted until the New Year. It was to be my first Christmas away from home, which was no hardship, as Evelyn, Hilary, young Andrew and I all had Christmas lunch with the Griffiths family in Buckingham. They owned a stationer's shop and were friends of Evelyn's.

After Christmas, I took some leave and went first to stay with Winifred and her mother and grandmother in Edinburgh. Then I spent a few days at Pineridge seeing friends and riding when I could.

Early in February I was shaken by the news that my department, Hut 10, at Bletchley Park was to move back to its headquarters in London. I enjoyed the work and the responsibility but I was happy in Adstock and I did not want to live in London. This was another turning point in my life. The decision had to be mine. The situation was made easier by an offer of a job in a different department based in the main house at BP. It was gratifying that my employers were reluctant to let me go — but I had made up my mind. On 26 February 1944 I moved into a department consisting of only three people — a Professor whom I had known since I stayed with his family in the Cambridge days, a Miss Mortimer, and me.

It was sad to see my friends depart. Evelyn wrote a silly verse to mark the occasion:

Lament

What shall we do now that HUT 10 is going,
Where will the cream of the Park be found,
How shall we know who is really worth knowing,
Will our position be socially sound?
Give us we beg a momentary respite
Do not immediately force them to quit.

What is the work that they do or they don't do?
What does it matter to us who remain?
They are the top drawer, the best people who
Set us a standard hard to explain.

Rarely they see us, NEVER they speak to us
Aloof and remote they lead their own life
Quietly conventional, deportment so decorous,
Oases of poise in a Park full of strife.

How shall we know if our conduct is worthy,
Who'll look down noses and see us below?
BP will be, and that very shortly,
Lower than ever and by low I mean low.

Please Mr Travis, oh please Miss Montgomery
Spare us these paragons, preserve our bon ton,
Don't send away our incredible nunnery, -
Who SHALL we laugh at when once they are gone?

The new job was day shift only. It was difficult, and I did not like it much, but there were always the stables to come back to and Evelyn and Hilary around for company.

We were given leave for VE Day on 8 May 1945. Part of the two days was spent distempering the stable with Winifred. From that moment on we started talking about the possibility of going abroad. An office was opened at BP where people like us (we were temporary civil servants) could discuss future employment with the Foreign Office. Somehow I managed to persuade Winifred that Jugoslavia would be the most interesting and exciting country for a posting. My holiday there in 1937 had been so wonderful that I had always wanted to go back.

On 1 June we started to take lessons in Serbo-Croat, three times a week from an Army officer at BP. The grammar is very difficult and complicated. There is also the hardship of learning the cyrillic alphabet. We worked hard at it, and at least made a beginning. Our names were down for the earliest posting to Jugoslavia with a note that we were learning the language. Not much hope was held out — and even less that we might go anywhere together.

VJ day on 15 August 1945 came and went. On 29 August, Winifred's birthday, we hitchhiked over to Whipsnade for the day. It was relatively easy and safe to hitchhike at that time — the lorry drivers being the best bet.

Then on 15 September we were suddenly summoned for an interview in London. Another interview followed two days later. Unbelievably we were both offered posts as archivists in the British Embassy in Belgrade. Little did we know in our innocence that archivists were filing clerks. It all seemed too good to be true. The pay, with foreign allowance, was better than we had been getting. On 5 October we left our jobs at BP giving ourselves a week to pack up at the stables. There followed the business of medicals and the necessary jabs, then a frustrating period of waiting.

On 19 November 1945 Winifred and I met in London and were driven from the Foreign Office to the dispersal centre in St James's Street. Our luggage was checked and put out on the pavement for transporting in a bus to the wartime airfield at Blackbushe. On arrival at Blackbushe we were assigned to an army hut near Hartley Witney. It was foggy and we were grounded. Identification of our luggage disclosed the awful fact that Winifred's suitcase was missing.

We had an uncomfortable night on tiered bunks in the hut with a bunch of very hearty Girl Guides. We were up at 4.30 am for an anticipated departure but it was still foggy. In desperation Winifred went by train to London to see if her suitcase could be traced at the point of departure from St James's Street. No luck — the suitcase never was recovered, but Winifred was given coupons to replace her clothes. Another night in our army quarters and we finally took off from Blackbushe at 11.15 am in misty conditions. It was my first flight, and I was decidedly scared. Over France the fog cleared and to my joy we had a good view of the Alps. After a four hour flight we landed at Istres near Marseilles and spent the night there — again in service quarters.

The following morning we flew on to Naples. Here there were more problems, and the plane to Bari did not arrive. We waited through lunch and tea and then were driven into Naples to the Hotel Sirena. Apparently we were without some necessary papers — which involved a visit to the Consul and a chance to see a little of the city. After dinner we thought we would take a walk, but were stopped by the hotel porter who said it was not safe. To confirm his advice — at that moment there was a great amount of scuffling, and a soldier came rushing in through the door followed by an Italian with a knife. Further scuffling in the hotel lobby ensued before the miscreants were evicted, and we decided not to go for a walk.

A Year in Belgrade

The year in Belgrade had a special quality. After the long dreary war years in England, a whole new field of adventure lay before me.

On the fifty-minute flight from Naples to Bari in a Dakota with bucket seats Winifred and I each spent half the time in the cockpit. The pilot and crew were South African, and so was I at heart. 'Great fun making the plane roll' I wrote home.

Even being stranded in Bari for a week had its moments. The Hotel Imperiale where we stayed as guests of the RAF was the height of glamour. There was a dance floor, waiters at our beck and call, and our own bathroom. And the water was hot. I hadn't had it so good for years — if ever.

First we were just waiting for our clearance papers from Belgrade, then the Jugoslavs banned all flights from Italy because dissidents had scrawled something rude on the aeroplane about Tito and the communists. So our stay in Bari was prolonged from day to day. The Consul obligingly advanced us some money and we dashed out to buy silk stockings — a great luxury.

We were 'adopted' by a South African army private and his pal, a sergeant in the cook house. They took us out in a boat and to the theatre and to the opera, which turned out sadly to be *The Geisha*. As we drove to collect some items from the NAAFI with our new-found friends, to our amazement we saw quick-fingered locals stealing valuables from open vehicles in slow-moving traffic. It was a good warning for future visits to Italy.

On 30 November 1945 we finally reached our destination and a cold (weather-wise) reception. So many things happened in those first few days that it was difficult to take them all in. The strange people in the Registry Office, the family on whom we were billeted, and the dreariness of the city all bore down on us like a heavy cloud.

In the Registry we had been greeted with 'We understand that you both speak fluent Serbo-Croat'. That was rather outfacing. We replied with gusts of laughter and then admitted we had had some lessons but could in no way be described as being 'fluent'. I am not sure at what stage we realised that we were just registry clerks. Archivist had sounded so grand and Jugoslavia so romantic — how could we be destined to be filing clerks? But even as filing clerks, our lives were scarcely ever dull.

The Embassy had only just ceased to be a Military Mission, of which Colonel Bill Deakin had been the head. He had been parachuted into Jugoslavia during the war. Now Ralph Stevenson was Ambassador, and we could not have had anyone nicer, kinder and more distinguished. He was one of the 'greats'.

Bill Deakin, a memorably brilliant and delightful person, stayed on for a time as Head of Chancery.

First we had to be installed in our billet, which turned out to be in a house with a family of four: Mamma and Poppa who spoke only Serbian, daughter Kaja, eighteen, who spoke some French, and a son Koste, sixteen, who spoke some German. Their house was ten minutes' walk from the Embassy. We paid them rent and supplied them with army rations every two weeks at our own expense. This took up just about all of our foreign living allowance.

We had two rooms overlooking the street. Our bedroom was overfurnished in a heavily eastern style. Window panes were missing from the four pairs of double French windows, and the central heating did not work. The other room, rather large, was a dining room where in solitary state we ate meals cooked by our landlady, who was rather easily reduced to tears.

The army rations for the two of us consisted of:

6 large tins of dried milk
4 tins of condensed milk
1 bag of flour
1 bag of oatmeal or cornflour
1 packet of either dried egg or custard powder
1 6lb tin of corned beef
2 tins of meat and veg
2 tins of salmon
4 tins of potatoes, 2 tins of carrots, 1 tin of peas
2 tins of Spam
4lbs margarine
2lbs lard
4lbs sugar
2 large hunks of cheese and 2 large hunks of bacon.

Apart from that, we had local butter, fresh meat sometimes, 2 loaves of bread and vegetables delivered about every other day. It was nice to know that the family were doing quite well out of us.

Apart from the food there were goodies from the NAAFI once a month: unheard of things like LUX soap flakes and other cosmetic and toilet items which were not available in England.

Soon after our arrival it began to snow, and before long it lay 3 feet deep in the streets. This was something new! There were sleighs and sledges in the streets — which was romantic and exciting. Perhaps there was a magic in Belgrade after all. But with no windows and no heating I wrote home for knitted socks to wear in bed. The Embassy took pity on us (perhaps because we made such a fuss) and the maintenance staff spent days rushing round trying to mend the pipes for the central heating. When they had mended the pipes they found that the stove would not work. They brought another stove and discovered that the pipes did not fit. I suppose they must have got it all together in the end because I recorded that we received a coal ration from the Embassy.

Our plight became a *cause célèbre*. The entire Embassy staff seemed to be running round in circles trying to help with the heating and the windows. Finally, they removed some panes from the Embassy windows to put in our bedroom. As with all new staff members, we were invited to lunch with H.E. the Ambassador. His first question was: 'Have you any windows yet?'

Once the glass had been provided, it was inevitable that new frames would be required, and I made a special expedition during work hours to have the windows measured.

The stables at Adstock had prepared us in a minor way for the discomforts of cold, but at least there we had hot water. Here there was none, so washing was not much fun — and the cold was considerably more severe than in Buckinghamshire, with temperatures of minus 5°C to minus 11°C, with snow as well. We now looked back on the stables as the acme of comfort and wondered how on earth we could have imagined that we were being spartan when we lived there. A bath at the Embassy was an occasional privilege in the early days — perhaps once a week.

Communication with our 'family' took place in a mixture of

Right: His Excellency Sir Ralph Stevenson, British Ambassador, and American diplomat Richard C. Patterson, Belgrade, 1945.
Below left: Winifred and me on a country outing from Belgrade.
Below right: Mrs Froebelius, Winifred and me outside the hotel at Novi Sad with the Embassy Jeep.

Serbian, French and German. Communication with the outside world was unpredictable. There was supposed to be a mail plane three times a week but sometimes there would be a ten-day wait owing to weather or political troubles. At the office we listened for the sound of an aircraft — and when we heard it we knew there would be mail.

During an interview with Bill Deakin in the early days he had held out some hope that Winifred and I might become couriers travelling within Jugoslavia to the various consulates in Skoplje, Zagreb, etc. Alas, they cut down on staff, and the courier idea was scrapped. It was worse for Winifred than for me to find that we had such lowly jobs: she had an Oxford degree, but I had no qualifications. What is more, it was only on arrival that we discovered that our contracts stated that we had signed on for three years.

It was the beginning of the Christmas festive season, and there was a variety of social occasions. Dinner parties with Embassy 'high ups' often included foreign diplomats, so long dresses were necessary and a knowledge of French was useful. These took place in elegant households with servants. (It is worth recording that at all these parties H.E. Ralph Stevenson wore red socks with his dinner jacket. Red socks, I was to find out later, were one of Peter's few sartorial eccentricities.) Then there were parties given by the lower ranks of army employees — which were rowdier and more beery and included our fellow workers in the Registry.

In our billet one evening, with some trepidation, we entertained an extraordinary mixture of the whole lot. Luckily — or thanks to the booze — it went like a bomb. Whatever type of party it was, vast quantities of liquor were consumed. Winifred complained that I never seemed to have a hangover, while she often suffered. Certainly I thrived on it. We usually spent our Sundays in bed after coming in from late-night parties, but that was as much to keep warm as anything else.

Winifred and I swept through the two rooms of the Registry in the basement of the building like a wind of change. Winifred had new ideas about filing systems, both of us were determined to read all the papers and as a result we were always able to find at short notice the previous papers on any subject. This was not, however, the way it had been done before and tended to slow up the initial filing. The typist was on our side and no doubt we were extremely disruptive.

New Year followed Christmas in a haze of good living and

warmth of spirit. Our Serbian family really put themselves out for us — which was especially kind as the Orthodox church does not celebrate Christmas until 7 January.

On Christmas Eve we returned to find a little Christmas tree in our room, with a bottle of champagne and packets of sweets. We invited a fellow worker from the Registry to our lunch and the family did us proud. We began with sherry with our soup, followed by a bully beef soufflé which 'melted in the mouth and was unbelievably good' (remember, we were not long out of post-war still-rationed England). Then came turkey, with tinned peas, rice, and potatoes. The turkey was a present from 'Poppa', as was the champagne which he had kept in his cellar for seven years. The 'English Christmas pudding' was a rather sodden plain cake with a few raisins in it, but it was set alight with Rakia (local raw spirit). After that came a deliciously rich coffee cake and *pain d'épice*. What a meal!

And so throughout the week one party followed another. On New Year's Eve the party was in our rooms and we mixed everyone together, including the Serbian friends of the family. We danced to a radio which we had just bought for £8 and which was tuned in to the Forces Network. Kaja (who had won a scholarship to the Prague Conservatoire) played the piano. At midnight we sang Auld Lang Syne in traditional style. The British soldiers and their girls left between 3 and 4 am. Then the Serbs sang songs and played the guitar and we sang our respective National Anthems and songs which undoubtedly were politically risky. We reckoned that it had been a very good party.

On 7 January came the Orthodox Church Christmas. I wrote:

> 'Personally, I find one Christmas a year quite enough. But it was amusing to see the customs here. On Christmas Eve we walked to a wooded hillside outside Belgrade with Kaja, Koste and some of their friends to fetch branches of oak trees which they call "Badnjak". Before the war, the King's guards used to go out on their horses and fetch the *badnjak*, then brought it back all decorated and rode through the town before taking it to the Royal Palace, where they were given drinks. Needless to say, this no longer takes place; in fact it is forbidden to cut down the oak trees. Anyway we got several large branches which were decorated with streamers. We had a small bonfire, and returned in the dark. When we got back

the family all kissed, and the two branches were put in the hall and the sitting room. We had sandwiches and red wine and at about 8 pm we had supper with the family who must not eat meat on Christmas Eve. Then "Mamma" brought in a small light in oil which "Poppa" lit and which burns all through Christmas. After which the son must recite the Lord's Prayer, "Poppa" must burn some of the oak leaves and then take four nuts which he throws into the four corners of the room. These remain there until Twelfth Night. He should also throw wheat on the fire. We drank quantities of wine and liqueurs and gave them their presents. To "Poppa" a bottle of whisky, "Mamma" some soap, Kaja some powder and Koste some pencils and chewing gum and they all seemed much pleased. Today — the 7th, Christmas Day, they seem to spend visiting and eating. The first person who comes in the house, if he is a man, brings luck and he must burn more oak leaves and throw wheat on the fire and he gets a coin for luck for himself. One of their male cousins arrived at 9 o'clock and I was taken to see the ceremony in the kitchen.

'For lunch we had pig's foot jelly, a wonderful soufflé with peas and tomato sauce, followed by Sparma which is stuffed cabbage leaves. The cabbage is rather like Sauerkraut though sort of fermented (!) and the stuffing must be a mixture of beef and pork — very good anyway. After that we had roast suckling pig which appears on the table head and all. We had seen the poor little thing before it was killed and felt rather sorry for it, it was so young! After that we had Torte made with nuts — delicious! After which — well it was rather difficult to move! Most of the people at the office had also been celebrating, so there was quite a holiday atmosphere even if we did have to work'.

The letter to my mother next day continues:

'Mrs A (Mamma) — who has gall stones — seems to have collapsed altogether today. At least she has done the cooking but we haven't seen her. They seem to have given up the idea of keeping the central heating up to scratch and now have it going rather mildly which does no good at all, and they have the stove in their sitting room. This means we eat in there and

> they eat in the kitchen. We wouldn't mind if (a) it wasn't our coal and wood they are using and (b) if we weren't paying rent for a room which is too cold to use and (c) if their sitting room wasn't so very hot and stuffy and smelly!'.

It seemed hardly surprising that poor Mamma was rather under the weather on the day after their Christmas.

Three days later, on 11 January I wrote again to my mother:

> 'You ask if hot water is a difficulty. Well, we wash always in cold water and sometimes if it is very cold we just don't wash at all!! Then about once in 10 days we steal into the Embassy residence [which was in the same building as the offices] and have a bath in luxury. The baths are unofficial but not forbidden and one creeps along the passage feeling one's shoes are squeaking and that the Ambassador may appear round any corner. Laundry doesn't amount to much because it is never worthwhile putting on clean clothes unless one has a bath! . . .
>
> 'Our coal ran out several days ago, and since then we have been eating and sitting in the family room. The last few days it has been cold, with mist and fog and the temperature minus 11°C. The trouble about not having central heating is that going from this room into the rest of the house is like going out of doors. No wonder we don't wash!'.

The problem of cold featured very large in our lives. Even the Embassy ran short of fuel, so that we were cold at work, too. By 20 January the notorious icy Kosova wind was blowing. It rained and the rain froze as it fell, leaving the streets and pavements a nightmare of slippery sheets and mounds of ice. The wind blew the loose snow in swirls till it looked like grey sand, and only the hard ice was left. The inhabitants seemed to be used to it, but for us walking was a dangerous game. Little boys were skating in the streets and one day there were people ski-ing in the street next to the Consulate.

During January we acquired a Serbian teacher and went to her house twice a week. She had a permit to teach French, and hoped that she would not get into trouble for giving us Serbian lessons. We liked her immensely — she was quite young, very good looking and always most elegantly dressed. It is to her credit that she managed to teach us quite a lot. Her husband, who was devastatingly handsome and spoke excellent French, was a professor of electrical engineering at the University. We met them occasionally at parties and enjoyed their company very much. In June she was arrested and in July we heard that she had been released from prison after eighteen days. We never knew why it all happened, but realised that it was nothing to do with her connection with us. After her release she went off to Slovenia with her young son, and that was the end of our lessons. But by then we had acquired enough Serbian to get ourselves around — if not to have philosophical discussions.

Both Winifred and I had chilblains and caught terrible colds so I sent an SOS for vitamins. My mother despatched two kinds — one of which looked exactly like Pregnavite, so I gave them to Winifred, who in the middle of a Serbian lesson was seized with a terrible pain and took two hours to recover. Nevertheless we ended up with her on the Pregnavite and me on the vitamin C. We also bought some army blankets.

At the time of our arrival in Belgrade, Jugoslavia had just been declared a Republic. Our entry through the 'Iron Curtain' had not been without its problems. Now we would really learn about communism in action. The Jugoslavs were supposed to believe that everything that was not communist was fascist and that the English and Americans were fascists. This was drummed into them on the radio and in the papers.

Every week there were official Partisan Street Conferences to which householders were supposed to go and to report on any anti-communist activities that they knew of among their neighbours. It was not an uncommon sight to see small groups of people being herded down the street by officers of OZNA (the Secret Police). It was a sinister and frightening spectacle.

Anyone suspected of potential opposition to the régime could be arrested, usually in the middle of the night, and marched off to prison. They might be tried on trumped-up charges, as 'collaborators' or as 'war profiteers'. Most people tried in this way were sentenced either to death or to life imprisonment and were never seen again. Many Jugoslavs lived in perpetual fear and we found that people were chary of being seen talking to us in public because we were English.

In spite of all the anti-British and anti-American propaganda, the Government was nevertheless quite happy to take UNRRA (United Nations Relief and Rehabilitation Administration) supplies and materials from the West.

In the Embassy we worried about the official British government delegations. They only saw what they were meant to see. How could we in the Embassy with all our local knowledge and inside information persuade them that all was not as it looked on the surface?

An incident in our billet really brought home to us the fear in which many people lived. Winifred had stayed in the flat all day with a cold. Late in the afternoon when all the family were out, a caller came to the door and more or less forced his way in, claiming that he was a friend of Kaja's. When I arrived back from work I found Winifred and the young man in the sitting room. The atmosphere was rather strained and conversation was difficult. At one point we left him alone for a few minutes while we went to find some photographs to show him. Eventually the family returned and it turned out that he was a complete stranger. He left and Poppa called in OZNA.

There was a tremendous fuss about our having let him in. The family were absolutely petrified, as they thought that the man might have been an OZNA agent sent to plant a compromising document in the house which could then be 'found' by an OZNA search party. We were questioned about the man's every move and they were appalled to hear that he had been alone in the sitting room, even if only for a minute or two. They searched the room, looking everywhere where a piece of paper could be concealed — under the carpet, behind the curtains, tucked into a chair — but luckily they found nothing. They ate no dinner, and we suspected that they spent a sleepless night, expecting OZNA to come back at any moment. This event shook us considerably.

It was while we were in Jugoslavia that we learned of the USSR's plans for dominating Eastern Europe and then the World. The five-pointed star was said to represent the five continents that Russia meant to control: Europe/Asia, Africa, America, South America and Australia. Africa was high up on the list of areas to be infiltrated. The kind of activities that were taking place are those which we have seen in many other countries, including Russia itself. Suspicion was the order of life. It was new to us then, and we saw it happen on our doorstep.

The political scene was full of drama. There was the problem of Mihailovic, Tito's rival, who was finally arrested while we were in Belgrade in April 1946.

The Jugoslavs were fighting for Trieste, which was still in the hands of the Allies, and it remained a bone of contention for a long time because the Italians were also claiming it. In Serbo-Croat it is spelled

'Trst' and much of the time it was big propaganda news on the radio.

The Church, too, was in trouble. A distinguished Archbishop called Stepinac was suspected of having anti-communist sympathies and the party was out to get him. Their manoeuvres were closely watched by us in the Embassy. On 4 February I wrote to my mother:

> 'Kaja, who had a music scholarship to go to the Conservatoire in Prague (she is a pianist) has just been refused a passport because she isn't a communist. The poor girl is miserable. She says that neither she nor her father can get jobs here because they are not communists. What a country! Anyone who has any leanings towards communism ought to be sent here. They would soon be cured. The trials are the most incredible things. The courts are filled with party supporters who see to it that the sentence is severe enough so that a defence counsel is more or less useless anyway. If the case is tried and a certain penalty demanded, and then the communists don't consider it hard enough, the case is brought up again, and sometimes again, the sentence more severe each time. There is literally no justice. Tito, whom I haven't seen yet, is reported to have terrific and genuine personal charm. Things have got quite beyond him I think. He has more or less lost control but presumably nothing can be done. He is not in a position to do much because he doesn't even know everything that is going on and anyway couldn't replace any of his officials and he will hardly be turned out because they have made him too much of a figurehead'.

Things may have got slightly out of his control at that time, but I believe it was he who was largely responsible for keeping his country away from total domination by the USSR. The fact that he was able to do that and to unite so many Balkan states (even if in a somewhat uneasy unity) is to his credit.

It was March before we managed to travel outside Belgrade for the first time. The weather was getting warmer and a few of us went out in the big black Daimler, past the barrier on the road, where the driver had to show his papers, and on to Avala. Avala is the only hill of any respectable size for miles around. A winding road leads to the summit, and to the tomb of the Unknown Soldier with a monument by Mestrovič: a good, solid, harmless sort of object in black marble. The hill is wooded, half with pine and half with oak and beech. At that time of year the ground was blue with wild squills, like a carpet, and there were wild snowdrops and primroses, too. Even though the day was grey, it was sheer joy after the wintry streets of Belgrade.

Outings of this sort were limited by the transport available, because we could only go beyond the city limits with an Embassy permit-holding driver. When we first arrived the Embassy had only Jeeps, but later there were three Daimlers and one Jeep — which were in considerable demand as more Foreign Office staff came in to replace the service personnel from the Military Mission days.

During the early months of the year we went several times to the Opera, which we much enjoyed; occasionally to the cinema, where often we were nearly overcome by the garlic breath of the pushing crowds at the entrance; and on one memorable occasion to see a troupe of Russian folk dancers. Though the dancing was superb, the crowd trying to get in was terrifying. Winifred had the tickets and she and I became separated. By the time I was shot through an entrance door like a pea out of a pea shooter, I was unable to do anything except move with the crowd. A girl not far from me was being beaten up by the commissionaire for not having a ticket. But I escaped unnoticed and eventually found Winifred.

By March the changeover from Military Mission to Embassy was complete. New staff appeared in the Registry, not always to our liking. Frustration with our jobs often made us behave very badly. A new typist with a 'terrible giggle' provoked us on her first evening into pretending that we had gone mad. After that she seemed a little frightened of us, but we had hoped that if we looked as though we were showing the strain too much the Head of Registry might send us off on local leave.

Local leave was something to which we applied a great deal of thought. We were allowed one month in the year, expenses paid. I had plans for Dubrovnik, Trieste and Switzerland, though I knew I could not fit them all in as each would need two weeks. I was also trying to arrange to meet Petronilla in Switzerland for two weeks' climbing in August.

At Easter a weekend party from the Registry was organised, giving us four days away from the office. Novi Sad is a rather pleasant town on the north side of the Danube, with the hills of Frushka Gora opposite. This was duck-breeding country, and there were flocks of them everywhere with their fluffy ducklings running behind. In the woods there were forget-me-nots, periwinkles and other lovely wild flowers. But the company was not very inspiring, as we were under the wing of the Head of Registry and his wife. The weather was not very good, either, so it was not a wild success. When we returned on the Monday night we found that our landlady's family had not arrived back from their weekend away. Not only was the larder locked up, but the power was turned off, so we spent the next day and a half living off the remains of our Novi Sad picnic meals. We scrounged some rations and heated up tins of meat, greens and potatoes on the electric fire in the Embassy telephone exchange.

Rations were once again plentiful after a very thin period early in April when fresh meat was unobtainable and the British rations had not come through from Trieste. There had been a few days when we would have been glad of a square meal. The local people were undoubtedly having a difficult time. Many items such as butter were obtainable only on the black market — and that was at a tremendous price as well as illegal. Our situation was saved when a lorry came through Trieste with 27 tons of food.

At the beginning of May, Winifred and I had had permission to go on leave together, but by the end of the month someone had realised that we could not both be spared from the Registry at the same time. On 31 May there was an opportunity for one of us to fly to Split in the Embassy plane. We tossed, and I won. The RAF Anson took us to Zemunik which is near Zadar, north of Šibenik, and once again I was invited to the cockpit and allowed to fly the aeroplane and even to remain in the cockpit when we landed. From there we went by ferry and Jeep to Split: a journey which took all day.

On holiday with me was Frances Coulson, the Ambassador's secretary, which made life easier because not only was she a delightful and amusing companion but she also knew the British Consuls around the country. We stayed with the Consul in Split for a few days in a villa just on the edge of the town above the sea. During the war it had been Gestapo HQ, and though ugly outside it had an exotic bathroom and was very comfortable. How lovely to lie in bed until 9 or 10 am and to sunbathe and swim in the sea! Most exciting of all perhaps was the visit to Mestrovič's house where there were a number of his works, some in walnut, some in marble. Unfortunately for me, he was in Switzerland at the time, so we did not meet him.

Jugoslavia was rich in sculptors of distinction. We had previously had the good fortune to be taken to meet Rosandič in his house in Belgrade. He was a dear old man with a long beard and a black skull cap. He and his wife both spoke French. Their lovely villa was full of beautiful furniture which he had carved, and in his studio he showed us some of his many wood carvings. Some of his works, of which he showed us photographs, were on exhibition in London and other capital cities.

From Split we went on to Dubrovnik by boat, calling at Makarska on the way. Makarska had suffered considerably in the war, scarcely one house undamaged. It was amusing to see Podgora again — the place where I had spent such a glorious week in 1937. Curiously I experienced a similar sudden gale near there again. It brought waves sweeping up on to the boat deck where we were sitting and then died down just as abruptly within the hour.

The hotel in Dubrovnik where I had stayed in 1937 was completely gutted, but the walled city itself was untouched and was as exquisitely beautiful as ever. And this time there were no tourists. We stayed in the Hotel Argentina: which had been, and indeed still was, one of the 'smart' establishments. Steps led down to the sea through a garden with a profusion of bougainvilia and oleander. We lay on the rocks sunbathing and eating cherries, swam in the clear blue water, and rejoiced in the idle life.

Having discovered that there was no hope of a lift back to Belgrade, we wired the Embassy asking if they could send a Jeep. We had hoped that the UNRRA representative might have been able to help.

We waited several days for a reply to our telegram. Finally, the Consul in Sarajevo rang up and said that he had been told to come and fetch us, but that he could not come before the King's birthday, and even then he would need a day in Dubrovnik. This gave us ten days in Dubrovnik — bathing, sunbathing and enjoying the wonderful fruit which we bought in the market. Cherries and apricots were in season, and peaches and plums just beginning. And so our leave stretched from ten to nineteen days.

The Consul, Dennis Wright, who was later to become Second Secretary at the Embassy, brought the British Press Officer (Mrs Rudoj) and also a girl from UNRRA, so we made up a happy party driving back to Sarajevo in his estate car. The drive was beautiful, first along the coast, then through bleak and rocky mountains, then for miles along a deep wide gorge where the Neretva river runs, then through 'Austrian-type' country. The further we drove into Bosnia the greener everything became. In Bosnia the corn was still green, while in Dalmatia it was already cut. The drive lasted from 6 am to

A Bosnian peasant, taken on the picnic lunch near Sarajevo.

5 pm with a stop at Mostar, famous for its bridge over the Neretva. The lunch was memorable, with fresh pink trout from the river, and everything laid on very specially for the 'Britanski Konsul'.

At Sarajevo we stayed in a hotel but had all our meals with the Consul. The day after we arrived he took us away up into the mountains for an idyllic picnic lunch. It was a showery day, but during a sunny interval we climbed to the top of a wooded hill.

Only in Africa have I met dusty roads as bad as those in Jugoslavia at that time. Even though we were in a closed car from Dubrovnik to Sarajevo we were coated with dust: even eyelashes and hair turned white.

At 6 am the next day a Jeep collected us for the drive to Belgrade up through the mountains in the early morning mists. Once over the pass, where it was bitterly cold in the open Jeep, we stopped at Vlastenica and fortified ourselves with rakia, after which we felt definitely warmer and better:

> 'We arrived in Belgrade at 5.15 pm — very dusty, bright red in the face from sun and wind, and very sore behind. The roads are incredibly bad. For miles one can't drive at more than 15 mph and even then one feels one is steeplechasing. It is depressing to be back in this horrid town again'.

Belgrade was undoubtedly a difficult and depressing place to live in during those days and at that time of year. The good news on our return was that Frances Coulson could pass on her flat in Resarska Street to us when she departed at the end of June. It would be sad to lose Frances but we were desperate to acquire a place of our own and get out of our digs. Almost immediately afterwards, Winifred left for Dubrovnik and it was my turn to do both her job and mine in the office.

We gave in our notice to our landlady who was pleased to hear that she would be having another lodger from the Embassy. Then we went to interview our new landlady who seemed very pleasant and self-effacing. She had one unusual question: 'Do you have bogs where you are now, because if so you must have all your clothes disinfected.' It took me a few minutes to discover what she was talking about. Not long after our move to Resarska Street she came to us in a great state of excitement because the flat below ours was being disinfected for bugs.

Before we could move we first had to obtain a police permit, and then we had to register with the Concierge, who would undoubtedly be keeping an eye on who came and went to our premises. One snag about living in a big block of flats like this was that if we came in after 10 pm we had to pay 5 dinars. At our previous place 'Poppa' was the housemaster, so we all had keys.

I had to complete the move in Winifred's absence. In a temperature of around 96°F it was fairly traumatic, but we inherited Frances' delightful maid Sofia. She unpacked much of our luggage, brought me my breakfast in bed and was an excellent cook. We were also able to practise our Serbian with her.

Because it was very hot and stuffy, we worked mornings and evenings and had the afternoons free. Whenever possible, and if a Jeep was available, a party of us took picnic lunches to the banks of the Sava outside the city and swam in the river, which though muddy, was very agreeable. We lived on melons — pink water melons, yellow water melons and canteloupe. Peaches, too, were plentiful and cheap — huge, yellow and juicy. A quote from a letter to my mother reads: 'We keep a water melon in the office and eat pieces, washing our ears with the ends, spitting pips everywhere and making horrid noises every half hour or so'. One wonders why we were not sacked. We spent much of our time fighting the rules of the 'establishment'. When we were not allowed a day off for August Bank Holiday, we considered organising a strike or demonstration. These were all symptoms of the bad feeling between the Chancery, now headed by the Chargé d'Affaires, and the Registry. We may have played hard and behaved rather badly, but we also worked hard and often worked on Saturday mornings. But we hated Belgrade in the summer.

In July there was a fresh crisis over a plane from Bari which came in with 'Down with Tito' and other anti-partisan inscriptions written in cyrillic on the fuselage. At that time there were tricky diplomatic negotiations going on concerning the future of Trieste and this incident set everything back.

The RAF crew were kept shut in the plane in the gruelling heat for 24 hours before being allowed to go to the RAF villa under a kind of house arrest.

With one local leave behind me I was already planning my mountaineering holiday in Switzerland. Petronilla, bringing my ice axe and boots, was to meet me in Grindelwald. But the only dates which suited her coincided with the time that the Assistant Archivist was having his holiday. The Head Archivist had agreed that the typist from the Registry and I could go on 17 August, and a room had been booked for us in the Hotel Adler. But on 18 July I wrote: 'My Swiss leave has been stopped owing to the bloody-mindedness of the Chargé d'Affaires. Although there were three people away at once before, he has put his foot down this time and even demanded a written statement from the Archivist that if I go away as well as the Assistant Archivist and the typist, the work will not be slowed up or suffer in any way, which the Archivist is somewhat naturally not prepared to do as we already have more than enough work to do'.

The only consolation was that I took over the Assistant Archivist's job in his absence, which offered a little more variety than my normal routine. I felt badly about it though, because I sat at an exalted desk while Winifred was still filing. I felt that she should have had the job, not I, but no doubt my year's seniority in age counted for something in this case.

At the beginning of August, morale began to rise in anticipation of the arrival of the new Ambassador, Sir Charles Peake, and his wife. An Ambassador with a wife would be a new experience — and, as it turned out, a very agreeable one. They were a delightful couple and created an entirely new family atmosphere almost at once. Suddenly, there was a great increase in social activities. There were lunches at the Embassy entertaining British MP's, Jugoslav ministers and foreign diplomats. They provided opportunities for me to use my French as well as a pleasant change from life in the Registry.

In late August the temperature dropped suddenly after a week of 104°F in the shade, to cool wet weather, which for the time being put an end to our afternoon bathing picnics. My next period of local leave was now due, and the plan was to go to Trieste and Venice, but there was a problem over obtaining Italian visas because of the strained diplomatic relations between the two countries. This time I was travelling with one of the typists called Betty McKay, and we finally set off by overnight train to Ljubljana where we spent the night in a hotel and the next day took the bus to Trieste.

The holiday turned out to be quite an adventure. The visit to Trieste was not a success except from a shopping point of view. It was the British Army that really got us down. We were staying in the big army hotel (The Excelsior) just full of officers and their wives and noisy, spoilt brats — all rank-conscious and smug and complacent-looking beyond words. Elizabeth Everard, a friend who was working there as a driver, was kind and helpful and entertained us on the Sunday, but we terrified her by occasionally bursting into Serbian. Not knowing more than about five words of Italian, Serbian sprang to mind more readily but with the delicate political situation it was not a good idea. Trieste and the surrounding area were under Allied Military Government at the time. Worse still, I had one of the Partisan Tito songs on the brain. Most of the anti-Tito refugees known as 'Dissident Jugs' were in Italy. At this point I even felt homesick for Jugoslavia.

After three days in Trieste we went by bus to Venice and on arrival our first two hours were spent trying to arrange accommodation. By visiting the Town Major's office during the lunch hour we were able to pull a fast one on the duty corporal by telling him that we were entitled to military accommodation. That had been true for Trieste but not strictly true for Venice; although still occupied by the Allies it was not part of a military zone. The difference to us meant that one night and full board cost fifteen shillings and sixpence as opposed to the civilian charge of £5.

The Europa was a vast improvement on the Trieste hotel and we had a wonderful time. The shops were a sight for sore eyes with their silks and all the wonderful things one could not buy in Jugoslavia or even in England:

> 'We walked and walked, we got lost, we saw several churches and of course the Doge's Palace. We went out in a gondola for an hour and a half one evening after dinner, and on the afternoon of the second day we even went to the Lido. Yes, Venice was well worth seeing. St Mark's church thrilled me most with its lovely mosaics. The Doge's Palace looks like something out of a fairy tale, but I was rather overwhelmed by the Tintoretto paintings — they are so vast and so many of them with angels and people floating about in the sky! Venice at dawn the day we left was unspeakably beautiful'.

Two days were not really long enough, but we were heading for the mountains. It was a bad journey back to Trieste with a very early start to catch the ferry and rear seats in the bus, where once more we might have got into trouble for speaking Serbian — mostly because I swore so loudly going over the bumps that a fat Italian turned and said: 'You speak Serbian?' So we answered 'Malo' ('a little') and then had a conversation with him in Serbian. Whereupon other people began to ask if we were Serbian and the whole bus turned round to look at us. Betty and I were so horrified that with one voice we shouted: 'No, we are English' and the whole busload roared with laughter.

On arrival in Trieste we found that UNRRA were quite unable to tell us if they could give us a lift to Ljubljana. They were so vague that we had in the end to book a room in the hotel, cursing the while because we might have travelled on an afternoon bus from Venice.

I was determined to arrange an UNRRA lift because on the way out our 104km bus journey from Ljubljana to Trieste had taken six and a half hours. Also, the bus was tiny, and although we had seats there were about forty passengers, some of them standing on our feet most of the way. What is more, Betty got lice and was badly bitten. Admittedly we had swept through the Customs in grand style, and while everyone else had to open their cases, ours were just left on top of the bus — but that was because the Consul's driver from Ljubljana had whispered in the bus driver's ear that we were English.

By 6 pm on Friday the 13th we had heard nothing from UNRRA, so we went round to Thomas Cook's, who were thoroughly unhelpful and sent us off to some place which we couldn't find for ages. Eventually after walking for miles in the wrong direction we came upon it, and there, to our intense relief, was the albino bus conductor from our inward journey. We burst into fluent Serbian and reserved seats for our return by bus at 7 am. Nevertheless we decided to make one last bid for UNRRA, and while at the office we bumped into a Viennese whom I had met in Dubrovnik. He was waiting for his girl friend who was coming with the UNRRA courier from Ljubljana. So we hung around drinking beer until 9 pm, then returned to our room in despair. At 10 o'clock our friend Elizabeth Everard came unexpectedly to our room to say she was sorry that we were leaving. It then turned out that the courier had arrived and was waiting for us. So after all we had a comfortable journey back in an estate car. The moon was nearly full and it was rather exciting tearing through Zone A and Zone B in the dark. There were three controls: out of Zone A; into Zone B; and out of Zone B. I never imagined I would be so thrilled to get back into this country (Jugoslavia). We had no trouble at all with the Customs who didn't open anything. We left Trieste at 10.30 pm and arrived at Ljubljana at 12.30 — or 11.30 local time. We then tried the only three hotels and found we couldn't get in anywhere. The driver offered to

Right: Joza Jurkovič at Bohinsko Jezero.
Below: Ray Bales, Tom Mansel. 'To Phil, May this small greeting bring you and all your friends in the embassy the very best of luck in the years to come. From two of the Embassy staff, Tom and Tommy. May your life be as lovely and happy as you are, Ray'.

take us to Bled but unfortunately we had left all our dinars at the Consulate. So at 11.30, having been up since 5.30 am, with much fear and trepidation we went and banged on the Consul's door. He was really very nice, but on our way up he had told us that we were his 51st and 52nd visitors since his arrival (he had been Acting Consul for a month!) and we gathered that he was pretty sick of visitors. However, he took us in, his wife made up the beds, and he shuffled off saying firmly 'Breakfast at eight'.

From then on everything was wonderful. The Consul drove us to Bohinsko Jezero which is beyond Bled, and by the time we had paid him for driving us, which he evidently enjoyed, he had quite forgiven us for our late night arrival and we parted firm friends. Not only that, he more or less insisted that we should stay there for a week, thereby overstaying our leave by three days. On our way through at the beginning of our holiday he had taken one look at us and said we ought to go straight to the lakes area as it was much more restful and certainly by the time we had finished with Trieste and Venice we did need a rest.

Bled with its big lake had a holiday home for Tito and was a well known resort. Bohinsko Jezero (Jezero means lake) lies further north with a smaller lake in the foothills of the Julian mountains. I described it all in a letter to Evelyn:

> 'It is most tantalising to be here looking at the summit of Triglav and not having any suitable climbing clothes. If I had my boots I could do it in two days from here . . . But there are plenty of walks up into the hills, there is the lake down below the hotel, and there is absolutely wonderful mountain air. It is just like Austria — the houses with their white-washed walls, wooden balconies and red geraniums, the churches with their turnip spires, and the people who are blond and good-looking. The drive up from Ljubljana was lovely. They are just finishing the second crop of hay and the autumn crocuses are thick in the new-mown grass. The Sava which we know so intimately in Belgrade, flows through the lake at Bled and through this one, too, having its source just about a mile beyond the lake where it comes out of the mountain in a waterfall. Here, at Bohinsko, it is a lovely clear, ice cold stream full of trout and flowing over stones. In Belgrade it is thick, dirty water, warm like soup in the summer, flowing over mud. We walked all round the lake today, a distance of about 11 miles, bathing on the way. The

> bathing was too cold for my liking, breathtaking, but the sun is hot'.

In my efforts to get a closer view of Triglav we spent the first three days taking strenuous exercise. We also went to the source of the River Sava, which turned out to be further than we had expected. Poor Betty was game for almost anything but she was not a mountaineer, and I think it was probably a relief for her when I found that one of the Croatian guests at the hotel was interested in hill-climbing. I had seen him studying a local map with an old man who looked like a guide, and in my best Serbian I had asked if I might look, too. The result was that the guest, whose name was Josip Jurkovič, asked if I would like to go up one of the routes with him the next day. Betty was happy to have a rest, Joza had a nice kind face and I was prepared to take a calculated risk for the sake of a bit of a scramble and a closer view of 'my mountain'. So I wrote to Evelyn:

> 'At last I am happy! Yesterday I went climbing to a lake high up in the mountains and then further to a place where we got a wonderful view of Triglav. It wasn't really climbing because both of us only had shoes and though very steep in parts the path was on the whole easy, but it was TERRIFIC! I learned that Joza was a Professor at a University and that he was 37. I was pleased because he puffed much more than I did and laughed when he said that he was surprised that "such a little woman" could go so far. We arrived back at 6.30. The sole of my shoe was flapping but I was happy.'

The letter went on:

> 'Back in Belgrade, bloody Belgrade. The Croat Prof. left Bohinsko Jezero at 8 am with a broken heart and I left at 12 noon — also with a broken heart. He was so nice'.

He did come to visit me once in Belgrade afterwards but was very nervous about the concierge who might wish to know his name and why he was visiting English people. For several years afterwards I corresponded with him by postcard using my best cyrillic script.

It was easy to write bitter things about Belgrade in the summer when it was a hot, dusty and rather drab city with ugly red stars on every building. Winifred has happily reminded me of the more glamorous aspect in the winter with pairs of oxen drawing carts, the trees and streets deep in snow with lamps hanging from wires over the middle of the road, and lots of horse-drawn sleighs with jingling bells. I think that is the right way to remember it.

By the end of the summer there was already talk of what would happen to Temporary Civil Servants such as Winifred and me — more establishment staff were being drafted into the Embassy. Winifred, with her Oxford degree, could sit the Civil Service examination. My position was more precarious and we both started thinking of our futures. Winifred would sit the exam and I would seek another job abroad if possible.

Sir Charles Peake, the Ambassador, had made the necessary arrangements for us to be repatriated to England at the expense of the Foreign Office, in spite of our three-year contract. Winifred was to be transferred. I was to resign.

I handed in my resignation at the beginning of October and we celebrated in style, with sherry in the office in the morning and dinner out in the evening. I was feted with flowers and caviar. At the time we had in our flat a case of champagne which we felt was too good to drink. I wonder what happened to it! We also had 52 tins of steak and kidney pudding, issued with our rations — unopened and unwanted.

When they heard I was leaving the Embassy, UNRRA offered me a job in Jugoslavia, but I felt that I needed a break, and turned it down. By 11 October I was suffering a few pangs of regret. Not much had been happening. It was getting cold and we were still having trouble with the planes bringing our mail. Letters at that time were few and far between.

We had a second outing to Novi Sad, this time in the delightful company of Desmond O'Meara, the Honorary Attaché. After lunch we passed a vineyard where the grapes were being pressed, so we stopped and went in. Needless to say, we tasted everything that there was to taste, and we came away with our water-bottle filled with wine, and feeling very happy. We were amazed at the delightful friendliness. The manager got everything working for us to show how it was done, even though it was Sunday. In Novi Sad itself our little party caused such a sensation in the restaurant that by the time we left we were feeling quite embarrassed. People stopped and talked to us and others came up and shook us by the hand and hit us over the shoulders and bowed and scraped — all because we were English. An English person out in the country there is a rarity. It was fun and lovely to get outside Belgrade.

A major sadness about our impending departure was leaving our maid Sofia. We had become great friends. She thought Winifred and I were a big joke. She laughed every time she saw us and looked after us in a wonderful way. She liked to tell me that I was 'weak' and that Winifred was so 'heavy' that she would wear her shoes out. We were known to the Jugoslavs who worked for the Embassy as 'Velika i Mala' — Big and Small.

After a fairly quiet October the social life hotted up again in November. Winifred went on her long-awaited leave, following in my footsteps to Trieste and Venice but not, as I recollect, to Bohinsko Jezero and the mountains. People obviously thought that I would be lonely, so I was out to dinner nearly every night.

Some time during the latter part of 1946 Chibbi Sturrock had arrived by train from England. During the war she had been dropped by parachute into Jugoslavia, and spoke fluent Serbian. It was said that this time her mother would not allow her to come by air. She was attached to the Press Office, and one of her remits from the Ambassador was to improve social relations with the Jugoslavs. So it came about that I found myself included in a small group with her and one or two others, joining Mr Nesič and a colleague from the Ministry of Foreign Affairs for an evening party. I had previously met Mr Nesič at an Embassy dinner. He was our principal contact with the Ministry and may well have been the Foreign Secretary. We were taken by our Jugoslav friends to a café, only to find on arrival that it was closing down to be taken over by the State, so there was nothing to eat. However, they took us on to the Majestic Hotel, where we had an extremely good meal and danced until 3 am. Nesič then insisted that we must all go to his rooms to drink coffee. The party went on for the rest of the night in what appeared to be a sort of bed sitting room. We were told we could not leave before 5 am and I was finally escorted home at 7. My conclusion was that though I was in favour of promoting good relations, it was very hard work.

When I went in to work and described the night's adventures my boss in the Registry, Mr Froebelius, expressed his disapproval in forceful terms.

Winifred returned from Trieste with a shocking cold and took to her bed with the whisky bottle. Inevitably, I followed in her footsteps. Sofia got toothache, and after we had introduced her to whisky as a cure, we found the contents of the bottle diminishing rather rapidly. Winifred also brought me a beautiful fur coat which I had seen in Venice. I had asked her, if she had time, to buy it for £29. For some reason she had had problems at the frontier at night getting back into Jugoslavia and was detained in a cold building where she had been pleased to have my fur coat to sleep in.

To celebrate my last birthday in Belgrade, on 22 November 1946, Winifred and I had dinner in the flat: caviar — a huge pile of it — soup, goose, and torte, accompanied by champagne and brandy. As it was Saturday we were able to walk it off the next day.

Our time was running out. We planned to leave Belgrade on 5 December. The last week was one long round of farewell lunches, dinners and drinks parties, a flurry of packing and goodbyes. We lived on a perpetual high — or perhaps it was just alcohol.

The Orient Express was there at midnight waiting in the cold station. Desmond O'Meara, Winifred and I had booked sleeping berths. A large party came to see us off including Chibbi Sturrock and the Froebeliuses. There was some confusion about our berths, as Winifred found a burly Jugoslav recumbent on hers, but it was finally sorted out.

We arrived in Venice on Saturday 7 December with four days for sightseeing. As far as I was concerned the first day was counted out, as I was exhausted emotionally and physically as a result of the strenuous last week in Belgrade and the touching send off. The Consul entertained us and we dined twice in the Danielle. We 'did' all the sights, enjoying being able to share the delights with Desmond, who had not been there before.

It was during this visit that, to our horror, the notorious floods occurred. As we hurried back to the Europa we could see people being helped from the rushing waters in St Mark's Square. Later, as we sat on stools at the hotel bar the water rose two feet and we had to make quickly for higher levels. Much has been written about these floods since then, but to have been there was a salutary and unforgettable experience.

We left Venice, again on the Orient Express, at 5 am on Wednesday 11 December, and arrived in Paris the following morning at the more reasonable hour of 9 am. After making ourselves known at the British Embassy I took charge of our threesome as tour leader, being the only one who had been there before.

During the next four days we made an excursion to Versailles, visited the Sacré Coeur, Napoleon's tomb at Les Invalides, the Musée Rodin, the Louvre and the Sainte Chapelle (alas! it was raining and the stained glass windows were not at their best). We also had lunch in the Champs Elysées. I took Winifred and Desmond to the Conciergerie because on a previous visit I had, in a rather macabre way, enjoyed the underground vaults. But they did not enjoy it much, and it also rained. However, on the Sunday we went to Mass in Notre Dâme and to the Opéra to see *Boris Goudenoff*, and that was a much happier day.

The next morning we left Paris, had a rough Channel crossing and arrived at Victoria at 9 pm. I was bringing home a large rug of traditional Jugoslav design from Belgrade for Evelyn and Hilary and had to collect it two days later from Customs, where I had to pay duty.

Winifred, Desmond and I had one more reunion at Winifred's mother's house in Kensington and Winifred came once to dinner at Highgate before I finally returned home to Droxford on Friday 20 December.

What I have not mentioned is that when we arrived in Belgrade Winifred decided to use her second name, Catherine. I gradually became used to it, but although I called her Catherine I always referred to her in my letters and my diary as Winifred. From now on she will be Catherine.

Pencil drawing of me made by Peter at Edwardes Square, 1948.

CHAPTER TEN

'Who's Peter Scott?'

The return journey from Belgrade at the end of 1946 was like the crest of a wave. The next twelve months were like a back wash from that wave. I had not realised how difficult it would be to get a job without qualifications. I wanted to travel. I had glowing references both from Bletchley and from the Ambassador in Belgrade. Without a degree I could not take the entrance examination for the Foreign Office. Catherine could and did, and went off to Trieste.

Applications to the British Council, UNRRA and other organisations occasionally led to interviews but got me nowhere. It was very depressing. We returned to an unusually cold winter. Food, fuel and many goods were still rationed in Britain. Post-war depression had set in. My friends were dispersed. Most of them were married or working elsewhere. Whisky was dead, and there was nothing to make me want to stay at home.

The news of Whisky's death had reached me in a letter from my mother while I was in Belgrade. It was a tragic story and I was miserable about it. He had stayed in Droxford with Barbara Wade, whose horses had the reputation for jumping out of fields. Whisky had never done that with me. But one night he got out with the others. In the morning they all turned up except Whisky. It was more than 24 hours before he was found dead in an isolated pond in a wood. It was an artificial pond with steep sides, so he could not get out. Even now I can hardly bear to think of him having such a terrible slow death by drowning.

I had the use of my mother's car. My cousin Alathea, widowed during the war, was at her old home near Bramdean and we met often. The summer came and went — with tennis parties, cocktail parties, periodic escapes to London to see old friends — but I never gave up hope of finding a job. My mother persuaded me to be secretary to the local Conservative club. The only good thing about that was that I learned a little about writing minutes of meetings — which later turned out to be useful.

A kind friend, Jane Chrystal, with whom I had ridden to Polperro during the war, let me ride and hunt her horse. Evelyn and Hilary used me as a child-minder: Andrew came to stay with me aged 2½ and I enjoyed the aunt/nephew relationship although I had always professed not to be good with children, and indeed did not particularly like children in general. Andrew was a 'deluthering' child — his speech sounded rather like that word. 'Itha bitha, itha bitha---', the non-words poured out of him in an excited stream. At an early age he learned his letters with delight but embarrassed me once in a railway carriage full of soldiers by reading out letter for letter the four-letter words written as graffiti on the mirrors which in those days were placed above the heads of passengers. 'What does that spell, Aunt Phoo?' 'Look at the cows', I said optimistically.

In September I went to stay with Hilary in Highgate. She was pregnant and I was to help with the children. In the event, on the day I arrived she went into the nursing home to give birth to Celia, and I found myself with Andrew and Hannah on my hands. Luckily Hilary's parents lived close by and came to the rescue so that I was able to go home three days later.

It was near the end of November that Antony Part rang me up one evening and said: 'Are you still looking for a job? How would you like to work for Peter Scott?' 'Who is he?' I asked. 'The man who paints ducks. You must have seen his pictures in the print shops.' I had forgotten about the handsome young naval officer who had made a fool of Kit and me at the party during the war when he pretended to be Jane Howard's deaf uncle.

Antony knew Peter's secretary, Elizabeth Adams, who had just become engaged to Peter's brother, Wayland Young. Liz Adams' stepmother Pamela had been with us several years before on one of our Austrian holidays, so there was a vague connection.

Liz first interviewed me over coffee in a tea shop in Sloane Street. She explained the job and must have thought that I would do. The next step was the interview with Peter, on 25 November.

Most of the self-confidence which I had acquired in Belgrade had slowly evaporated during the long months with their sterile interviews. Nevertheless I had a determination to try anything. But I was nervous. I arrived with my two glowing references. We sat at the long refectory dining-room table and talked of this and that. No mention was made of references. How soon could I start? For some reason 10 December was chosen, and I arranged to stay with Aunt Dora Backhouse in London until I could find digs.

I was to be Peter's personal secretary, and Assistant Secretary of the Severn Wildfowl Trust. The Trust was just one year old and had only about 500 members, so it had little money, and I was paid by Peter. The office was the dining room at 8 Edwardes Square in Kensington, which was Peter's house. At one end of the room were the refectory table and chairs. French windows led out into a small garden. At the other end was Peter's desk, a small desk for me, and behind.me against the other wall was a table with a typewriter for Douglas Eccleston. Douglas had been at Bletchley Park and after leaving had gone directly to work for Peter. He was the most brilliant typist I have ever known and also one of the most delightful people to work with.

Occasionally the office overflowed into the upstairs drawing room. This happened when we were sending out circulars to the SWT members, and I roped in friends to help put them in envelopes. Peter's five year-old daughter Nicola and her Nannie were living in the house, and Nannie (Buss) was very put out about using the drawing-room as an office. She did not approve at all, and her mouth turned down in an inverted U. Dougie made wonderful jokes about it and she relented a little — after all, she had known him longer than me, but she still did not think it was the right way to treat a respectable house. I was rather frightened of her at that time.

Space was very limited in the dining room with three of us — sometimes four when Ray Gregorson (Aickman) joined us as organising secretary for three days a week. My desk was so small that I often put my filing on the floor. If I was foolish enough to leave it there overnight, there was the risk that Nannie Buss would have told Joan, the coloured living-in cook, to throw it away. Dougie and I would arrive at 9 am never knowing quite what to expect. If we could not find papers when Peter asked for them we suggested that perhaps Nannie had thrown them away.

She finally came to accept me when I offered to look after Nicky on her day off, if her mother Jane could not have her. The fact that Nicky quite obviously liked me helped. Too often Nannie's days off included interesting assignations with men. But they seldom, if ever, turned up. It was rather sad. She had a curious habit with men of staring at their fly buttons as she talked. Keith Shackleton used to find this very disconcerting.

As well as Peter, Nicky, Nannie, Dougie, Joan and me, there was Bushy. Bushy was a large white Pekinese with great character. From time to time he would enter the dining room in the middle of the morning in order to cause a diversion. There would be cries from upstairs of 'Bushy, Bushy, Bushy . . .' But Dougie was obviously more fun than Nannie, and Bushy turned a deaf ear and ran riot. He used to go quite mad, tearing round in circles, leaping on to Dougie's table, scattering letters and papers everywhere. It was rather like having a large snowball whirling round amid the snowflakes. We loved him and forgave him the chaos for the hysterical laughter he always provoked.

When I started work, Field Marshal Lord Alanbrooke was President of the Severn Wildfowl Trust, Sir Archibald Jamieson, Chairman of Vickers, was Treasurer, and Michael Bratby, a stockbroker, was Honorary Secretary. Michael was an old Cambridge and wildfowling friend of Peter's. I was to see a great deal of him in the next few years. He died of lung cancer in 1959. His witticisms have passed into our everyday language. He was known to a small group of friends as 'the Major' while Peter was 'the Commander'.

Another special friend of Peter's who was also often in and out of the house was James Robertson Justice (one of the founder members of the Council of the Wildfowl Trust). Not long after I started work I came in one morning to find Peter sitting in his dressing gown (as he often did) finishing his breakfast, and with him was this huge lovely man with twinkling eyes who had come to inspect the new secretary. Alas, he is no longer with us, but he was a treasured friend to us both. In later years when he was already ill we spent a happy week staying with him and his lovely Irina at Spinningdale in Sutherland. We had the children with us and I was afraid that he would find them a nuisance, but he told me he had fallen in love with Dafila.

James was a great wildfowler. On one occasion he arrived at Edwardes Square with some Pinkfooted Geese which he proposed to cook in his own special way. He took over the kitchen and left them simmering in red wine garnished with other bits and pieces. Joan, the cook, was not too pleased about the clearing up afterwards but we had all taken some part in the preparations and I think she really enjoyed it.

Some time in January I moved into digs in Courtfield Gardens, off Cromwell Road. The room was typical of any other single owner bed-sitter in London. There was a gas fire, an electric ring and sparse furnishing. The room was three floors up and the telephone was on the ground floor. The bathroom, which I shared with unknown

Nicky, aged four, with her white Pekinese, Bushy.

characters whom I never saw, was unattractive. I learned the art of getting completely dry before leaving the bath, in order to step straight into slippers without touching the floor.

It is surprising that I survived the initiation period as a private secretary. Although engaged as Assistant Secretary to the Severn Wildfowl Trust I was also Peter's Personal Assistant. At that time he *was* the Severn Wildfowl Trust. Not only was he paying me, he was also paying the salary of the Curator who at that time was Eunice Overend. She had succeeded John Yealland who was later to return to the Trust. I had no secretarial experience for this kind of work. I could type with two fingers, but with Dougie to take shorthand and type it was not necessary.

My shyness and inexperience made me appear abrupt on the telephone, and one of Peter's friends complained that I was too fierce. It could only have been fate and Peter's kindliness that kept me there. At that time he was a lonely person. His first marriage had just broken up and his mother had died, all within a year. I was the third Assistant Secretary of the Trust, and he probably did not want further changes going on around him. His personal finances were in a mess. Accounts were not my thing and money was not his thing.

Before long, Ray Aickman, wife of the founder of the Inland Waterways Association, was engaged to come two days a week to do the Trust accounts and to help generally with the Trust work. But Peter's private accounts were left to Dougie and me. Once a month we went by bus to Leinster Corner taking the ledger to Peter's step-father, Lord Kennet, who would go through it carefully while Dougie and I sat in trembling silence. One day he suddenly turned to Dougie and barked 'What is this payment to Miss Overend which keeps coming up?' Dougie was too frightened to speak, and I explained that this was the Wildfowl Trust Curator's salary. We did wonder what he thought, and we laughed all the way back to Edwardes Square.

Peter and I had a certain empathy for each other from the start, otherwise I should never have held the job down. But the spirit of independence still burned strongly, and I needed a life of my own outside working hours. Salvation from the dreary bed-sitter in Courtfield Gardens came early in March 1948 when Catherine returned from foreign parts and we found a flat at 32 Ladbroke Grove where we set up house together. The house belonged to a distinguished pianist called Mr Kitchen, who was abroad. We had the basement with garden and the second floor. Otherwise the house was empty. Catherine was still with the Foreign Office. So once again domestic chores could be shared in a light-hearted manner and, as before, we laughed and enjoyed our free time together with no feeling of dependence on, or obligation to each other. Having a bit of garden was bliss.

Peter's friends, too, began to play an important part in my life. Keith Shackleton and Michael Bratby were the most frequent visitors. Michael was perhaps one of Peter's closest friends, and these two were destined to be godfathers to our daughter Dafila. Michael had been a wildfowling friend. Keith, much younger, had originally approached Peter for advice about his wildlife paintings.

My weekends were spent either at Droxford, where I could still borrow a horse sometimes and hunt, or at the New Grounds learning about waterfowl and the people who worked there. There were nearly always other visitors to Slimbridge. Peter was learning to fly, and offered to pay for lessons for me. Kidlington, near Oxford, was on the way to Slimbridge, so we stopped and had our lessons there when the occasion permitted. Keith, too, was learning to fly, though his base was Fairoaks in Surrey, where I also once had a lesson. For some reason Keith adopted the name of the Chinese Aviator — Wun Wing Lo. Postcards and messages from him always came written in the Chinese vertical style with the letter L substituted for R. Peter and Keith had both been learning for longer than I, and duly achieved their licences while I had only put in about six hours' instruction. One day, Peter offered to fly me over the New Grounds in a Tiger Moth. It was not far as the goose flies from Kidlington. We looped the loop twice over the Dumbles, which I found both scary and exhilarating. Then we flew over the village. Later the vicar complained that we had been doing aerobatics over the church spire in a dangerous manner, which was quite untrue.

Learning the names of all the birds took time — and there was more to learn than that. Keith told me that if you spotted a Lesser White-fronted Goose you got a medal and if you put up the wild geese you were in deadly trouble. In those days we had to crawl bent double behind the sea wall to get to the little straw hides which had been built by German prisoners-of-war. I had to learn about sitting in cold hides for hours looking for rare geese without making a sound. It was reported that Dougie had nearly died of exposure during his first introduction to goose-watching. For some reason Keith always

With Eunice Overend and Peter, looking at black swans in a newly converted enclosure at the New Grounds, 1948.

got the blame for putting up any of the wild birds and he acquired the title of 'The Scapegoat of the Trust', a title which he bore with great good humour. He was kind enough to christen me 'The Toast of the Trust' which shortened simply became 'The Toast'.

There were some interesting characters among the staff at the New Grounds. There was the Keeper, Mr Cameron, who lived in the bungalow. He and his wife were Scottish. The half-mile bicycle ride to the pub by the canal meant nothing to 'The Old Mon' as we used to call him. Coming back was more precarious, of course, but the road is straight. His wife cooked lovely honest, good, country meals for us in the cottage, kept the Rayburn cooker burning, put hot water bottles in our beds, and mothered us just as much as we wanted but no more. Her drop scones, or pancakes, were famous. She was to work full time for us in the cottage after Peter moved out of London.

They had two pretty daughters. The older one, Betty, was married early but the younger red-haired Peggy started work for the Trust at seventeen. Tall, slim and boyish, with green eyes and a flaming mop of hair, she was very attractive. All the boys were in love with her. She married some years later but stayed on to work with the birds. Sadly, the marriage seemed to be a disaster. She finally fled, never to be seen again. Some valuable tools she had borrowed together with all her widowed mother's savings disappeared at the same time.

There was Mervyn Everett, allegedly poacher turned keeper. Dark and swarthy like a gypsy, with colourful language and a splendid sense of humour, he was an excellent ornithologist. He sometimes worked the duck decoy. 'That bird fled down the pipe and ketched itself' he used to say. It was a pity that another member of staff saw him emerging from the decoy with some mallards tucked under his coat. He had to leave and it was a great loss.

There was Tommy Johnstone, who had quit medicine for the birds. When I arrived in 1947 he had not been with the Trust for many months and was working as a warden under Eunice Overend. He and his beautiful wife Diana lived in Yew Tree Cottage in Slimbridge with their small daughter, Carol. They were to become special friends of ours, and Carol was married from our house. Later Tommy became Curator at Slimbridge and Diana ran the gatehouse and shop, as well as doing much for the general public relations of the Trust and Peter.

Back to the Mountains

In 1948 the Olympic Games were held in Britain. The yachting events were at Torquay. Peter was on the Council of the Yacht Racing Association, and as Chairman of one of the committees he was very much involved with the arrangements. I had a terrible time with the mass of paper coming through. I knew nothing about yacht racing and had difficulty keeping the filing in order between the YRA and the international body, the IYRU. Even Peter became slightly impatient with me and it was with some relief that I saw him off to Torquay and ventured forth on a holiday.

The mountains had called since before the war. No-one could come with me, so I must go alone. I went to Cook's and for some reason decided to make for Megève which is now a popular French ski resort. It seemed to be near enough to Chamonix for some climbing, and the hotel that I chose was sufficiently in the country to provide good walking prospects.

I was due to leave London on Saturday 31 July 1948. On Friday evening I discovered that my passport was at home in Droxford. I rang my mother who undertook to put it on a train the next morning if I met it at Waterloo. My train from Victoria was due to leave at 9.05 am that day, so I spent a worrying night. But the passport did arrive and I did get to Victoria in time, with a rucksack on my back.

The train journey was terrible: very hot and very crowded. Changing trains in Paris was not as easy as I had expected but I had a bunk for the night journey down to Lyons. The bus trip from there to Megève was a nightmare. I stood most of the way. My ankles swelled up like balloons and remained like that for twenty-four hours.

The Hotel Mont Blanc was quite small and the people were friendly. For the first few days I explored locally. It was disappointing not to be able to see Mont Blanc, and the weather turned nasty, with a thunderstorm and rain. Two teenage Algerian boys came for walks with me and I had a pleasant little climb up Mont Joly with a delightful Italian called Ermanno Maccario from Toulon who was staying in the hotel with his wife. We came down with the télé-

Right: In the téléférique near Megève.

Left: View of the French Alps taken from the glacier above Chamonix on the way to the Col du Géant.

férique, and on our return to the hotel at 7.30 in the evening I was greeted with cries of 'Comme vous-êtes solide!'. It had been a steep scramble — nothing more — but from the top I had seen the snow-peaked mountains.

So now I had a week left for climbing, and the weather was dicey. By the Friday of the first week I had made enquiries and had booked a guide for 2 pm on the following day. But it rained and I did not go. It was Monday before the opportunity came. It had rained all Sunday so I had not planned for the next day. I woke to a clear blue sky, but I could not believe that it would last and so missed the only bus to Chamonix at 9.50 am. By then I had decided it was now or never, so I went all round the town enquiring about means of getting to Chamonix. I was told that it was impossible until the evening. Further research and questioning revealed that by changing buses twice I could get there by 3 pm. So I rang the Bureau des Guides and made a date. Then the hotel waiter said there was a courier bus that left at 11.30, so I rushed to the café only to be told that the bus had left at 10.30 — and by then it was 11 am. As I came out of the café I saw a bus in the square with a driver standing by it. I asked if he was going to Chamonix and he said 'Yes' — but it was an excursion from Annecy and he was full up. I pleaded with him, and finally he said he would take me in ten minutes. I changed into climbing clothes, collected some bread, cheese, ham, eggs, fruit, etc, from the hotel and ran. I found a seat in the bus and arrived at Chamonix at 12.30. At the Bureau des Guides they said I should be ready to leave at 2 pm. So I sat on a seat in the nearby park and ate bread and cheese.

We took the funicular train from Chamonix to the Mer de Glace station — just my guide, Charles Kleber, and I. From there we climbed up the glacier and round into the mountains. At about 5 pm we arrived at the Réfuge du Réquin, which is a hut perched on a rock. There were something like 120 people there, many waiting to climb Mont Blanc, but we were aiming for the Col du Géant. We slept in dormitories. I went to bed at 8.30, lights were out at 9 pm, and I heard it raining hard for about three hours. At 3 am one of the parties was roused, so that was the end of sleep. At 4 am my guide said 'Allons', so I got up. We started off eventually at about 4.45 under an overcast sky. We roped up on the glacier, where the guide had to cut steps in the ice. Going over the *seracs* was something new and different in my climbing experience. After that we crossed a snow plateau — and then climbed — it seemed endlessly — in the snow. It was steep and I puffed like mad. By this time it was snowing — a blizzard — and it was cold. Eventually we reached the Col du Géant, which is about 10,500 feet and overlooks Italy. But we could not see a thing. So on to the Réfuge Torino which is down a bit on the other side and actually in Italy. There, five English people who had been at the Réfuge du Réquin, and two other parties, turned up including a party with a Curé. It made me laugh to see him lifting his long, black skirts and jumping the crevasses. We stayed there for an hour and a half and had breakfast and hot grog. My trousers were so wet that I could hardly bear to sit down and I shoved a sweater inside my trousers to keep all the damp clothing off me. It was cold, too. I had an immense hunger and ate quantities of my by-then stale bread, and one of the English party stood me a bowl of hot chocolate because I was so nervous of buying anything in case I could not pay the guide at the end of it all. I had to pay for his meals as well as the tariff, which was £5.

At 9 am we left the hut. As we came out the mist lifted for a moment, and miles below we saw the valley, with Courmayeur. Then it clouded again and as we came up over the col we were met by snow driving in our faces. Our previous tracks were obliterated by new snow and we could see nothing except whiteness everywhere. For about fifteen minutes we were lost and there was much shouting between my guide and some others behind who had followed us. We found our way and after half an hour suddenly it cleared so that there was sunlight on some of the peaks. But Mont Blanc never unveiled itself. We glissaded a lot of the way down, which was something I always enjoyed. Back at the hut someone whom I had spoken to the night before had left a message to say that if I was at the station in Chamonix by 3 pm he would give me a lift to Megève. So we went down the glacier, unroped now, and caught the funicular train at 1.15. I made my rendezvous successfully and got my lift by car.

It was a marvellous adventure. It might have been even better if the weather had been kinder, but I was happy. There was one day left to recover, then I made the uneventful journey home.

Peter feeding the ducks in the orchard, with the old duck house in the background, 1947. The Caribbean flamingoes are now in this pen.

The Severn Wildfowl Trust

It was a relief to me, as well as to Peter, that he decided during 1948 to come out of politics. When I first went to work with him he was the prospective Conservative parliamentary candidate for Wembley North. I like to think I had some influence on his decision. Anyone could be a politician but not anyone could paint birds like he did. And there were so many other things to do. Creating the Wildfowl Trust began to take up more and more time. Although Peter did not then own a boat he still took part in some dinghy races, and he always went to Cowes Week.

That first year, 1948, Nannie, Nicola and I stayed at a boarding house in Cowes while Peter and his friends were up the hill with June Damant, an old family friend. It was all very strange and new to me but it was fun, and in later years I too stayed in June's house, went to the Royal Yacht Squadron Ball, and even raced one year in a Swallow with Peter and John Winter. Usually, I was left ashore while Peter raced in someone's boat, but if the weather was too rough for racing, Peter would take me out and give me sailing lessons. Catamarans were just coming on the scene and the makers were anxious to influence Peter, as a Council member of the RYA, in their favour. The nearest I have ever been to capsizing was in a prototype catamaran on a very wild and windy day at Cowes. One of the lines got jammed in a cleat. I had no idea what would happen. When you are sitting on the side of a catamaran it looks a long way down to the water as she keels over. Peter said quite calmly 'I think we are going to capsize'. Just in time, the line freed and we sank back into the water.

I think I would have enjoyed sailing if I had learned early enough. As it was, opportunities were few and far between. The best time I had was crewing one year in a Jolly Boat at Cowes with a sympathetic helmsman. We did not win but it was blowing quite hard and it was exciting. The worst experience I had was in Norway many years later. We were staying with HM King Olav in his yacht for the big regatta near Oslo. Peter was racing with the King. The Commodore of the Yacht Club offered to take me as a crew member in his boat, an International One Design. Normally there is a crew of five, but he was racing with his wife and thirteen year-old son as crew. It was blowing hard. They shouted instructions to me which I did not understand; perhaps they were in Norwegian. Finally, they decided I was just in the way, so they sent me down into the cabin where I was very nearly seasick. In spite of that, they came in second or third. After that I was definitely turned off sailing. Better for me to stick to the birds.

It was early in 1948 that the first ever attempt at rocket-netting was made, and I was there. Catching wild geese is notoriously difficult (hence 'a wild goose chase'). Our purpose was to ring them for behaviour and population studies — and the idea of using rockets had been conceived by Peter and James Robertson Justice. Now here we were, ready to put the idea to the test. It was a big moment in Peter's life and held great promise for the future.

It was a 4 am start with a long walk through the fields, and it was cold. But it was worth it. The catch was 31 geese out of 1300 on the New Grounds at the time. It was the beginning of a long rocket-netting saga.

I was beginning to understand about the Severn Wildfowl Trust, and Ray Aickman and I had a good relationship in the office. Her particular forte was looking after the accounts. Later, Catherine and I had to leave Ladbroke Grove when the owners returned, and in the summer we set up house in a flat in Scarsdale Villas. It was closer to Edwardes Square but had no garden and was less attractive.

In October 1948 Peter went to America and I suddenly found myself running the Trust. To back me up I had Mike Bratby and not least Ray Aickman who later, as Ray Gregorson, was given the title of Organiser. Lord Alanbrooke was President. His great hobby was filming birds. He was one of the kindest people I have ever known and one of the most unassuming, and it was always a pleasure to organise his weekends at the cottage. He, Mike Bratby, Keith Shackleton and Gavin Maxwell would often come for the weekend, whether Peter was there or not.

At that time the Trust consisted of two large enclosures, the cottage, the bungalow, and just a small number of staff. There was always something for me to do, and I enjoyed the challenge and the fun. However, I did go home for occasional weekends.

'I suppose you are going to marry Peter Scott,' one of my riding friends said one day. 'Good Heavens no,' I replied. I was foot-loose and fancy-free and had a whole new world of friends.

One of the occasional visitors to the New Grounds was Ludwig Koch. He was collecting recordings of bird song with an apparatus which was amazing when contrasted with modern counterparts. Large and heavy, it needed a team of people to carry it around. In particular Ludwig wanted to record the sound of the wild White-fronted Geese. Elaborate arrangements were made to drive the geese over his temporary hide set up in a hedge. They duly flew over in a great flock. Everything in the apparatus seemed to be running. As the last liquid call died away, Peter asked: 'Did you get anything?'. 'I heard nozink', Ludwig replied, then smiled happily.

Peter was not at Slimbridge on the occasion when Ludwig fell down the stairs. He was sleeping in Peter's bedroom above the studio which had very steep stairs leading straight up into it. From the kitchen I heard a crash and a groan and I rushed out to find Ludwig lying at the bottom of the stairs. At that time he was probably in his seventies. I helped him into the studio but he was obviously very shaken, and sat there moaning. I thought about administering whisky or brandy, but decided that Mrs Cameron was the obvious person to call on for help. I hurried over to the bungalow where she and the 'Old Mon' lived, and she came over. Whisky did indeed prove to be the answer and he was soon restored.

The film *Scott of the Antarctic* was made in 1948, and from time to time Peter would go to Denham Studios. Diana Churchill, who was playing Kathleen Scott, came to Edwardes Square for advice. Once Peter took me with him to the studios and I enjoyed meeting some of the cast. But the biggest thrill was the Royal Command performance in London. Peter had been sent two tickets but was unable to get back from America in time. I sought advice from Mike Bratby. I thought he might go. He discussed it with various people, including James Justice who was playing the part of Petty Officer Evans in the film. He rang me back and said 'You and Ray can go'. What excitement! We had to wear our best evening clothes. Chokers were in fashion and Jane Howard made me a beautiful blue ribbon one embroidered with black beads. Our seats were not far from those of the King and Queen and other members of the Royal Family. It was unbelievably

Left: The cottage, with the studio window and the cow byre in 1949. The byre was later converted, by raising its roof, into hostel bedrooms.
Below: Keith Shackleton and Peter extricating White-fronted Geese from the rocket-net in the 100-acre at Slimbridge in 1948. It was the first-ever catch using this method.
Right: Stalking the Whitefronts behind the sea wall by the Dumbles, Slimbridge.
Below right: Ludwig Koch recording White-fronted Geese, 1949 . . . 'I hear nozink.'

thrilling for us. The film was a masterpiece and has been shown many times since.

Evelyn was living and working in Hamburg with the German division of BP. He and Hilary invited my mother and me to join them and their family for Christmas. We flew from Northolt which was still being used for European flights. Unfortunately, the weather prevented us from landing at Hamburg so we were diverted to Copenhagen. My mother, then sixty-nine, had never flown before. It was interesting to see some of the sights of Copenhagen but we were more concerned with getting to our proper destination in time for Christmas. The fog persisted, the airline abandoned us, and we finally managed to get on an overnight train to Hamburg. My overwhelming memory of that Christmas was being driven by Evelyn through the devastated areas of the city.

Early Days at Slimbridge

Back at Edwardes Square things were hotting up for the great move. Peter was more or less broke, much of the funding for the Severn Wildfowl Trust having come from his own pocket. Keeping the house in London as well as the weekend cottage at Slimbridge was too expensive. Bill Kennet, Peter's step-father, advised selling the house. So Number 8 Edwardes Square was sold for £8,000.

The job of cleaning out the cellar and packing up the files was left to Dougie and me. Joan the cook had departed, and a delightful friend of Jane Howard's called Shelley came to help with cooking and other chores. Nicky and Nannie were installed with a friend of Jane's in Evelyn Gardens.

Emptying the cellar was the most exacting task. There was a formidable array of tin trunks filled with papers, letters and naval uniforms, some of them dating back to the Scott expeditions. Decisions had to be made about what to take to the New Grounds, what to store, and what belonged to Jane.

The operation was completed on 14 January 1949. The furniture had gone into store the day before, and the huge van was loaded with items destined for the New Grounds. I was on hand to receive them at the other end: but it was not so simple. The canal bridge was in two parts which at that time met in the middle. A barge had hit one arm of the bridge, knocking it crooked on its pivot. I had no way of letting the removal men know. So the van duly arrived at the Patch bridge. At first it seemed possible that the bridge might be mended without delay — perhaps in a few hours. So we waited for the rest of the day, with telephone calls to the canal company and discussions across the water with the van driver and the bridge man.

Damage to the bridge was by no means a rare happening, but it could not have come at a more inconvenient time. There was a small ferry boat for foot passengers, but on this occasion it hardly met our requirements. Finally, the van driver insisted on taking his vehicle back to London. He was not allowed, he claimed, to leave his van unaccompanied overnight, and he did not intend to sleep in it. So back to London went all the furniture and books.

It was three days before the bridge was mended. The cost to Peter was £25 more than the straight journey down would have been. Our efforts to reclaim this from the canal company were unsuccessful.

Life at the New Grounds was increasingly busy. Dougie was housed in lodgings in the village and I had my own room in the three-bedroomed cottage. There was no electricity, but we kept warm with a Rayburn cooker in the kitchen, a small stove to heat the water in the passage, and an open fire in the studio. The Aladdin and Tilley lamps also helped to heat the studio at night.

There was never a dull moment: lectures to organise, a large mail to deal with, a variety of visitors to entertain, and the Trust outdoor activities. From time to time a Lady Malise Graham, who had looked after Peter as a boy, used to ring up and tell me how lucky I was to be Peter Scott's secretary. Up to my eyes in work and trying to keep Peter's appointments and lecture engagements in some sort of order, I did not always appreciate her remarks. Once, I plucked up courage and said to her: 'You should try it'. I am not sure if she approved of me.

In the May of 1949 Peter went on a long expedition to the Perry River in Arctic Canada to search for the breeding grounds of the Ross's goose. From the long letters that I wrote to him during his absence that summer, it is evident that I was already falling under his spell. Events at Slimbridge, the breeding success of the birds, new arrivals, the problem of the proposed bombing range — all were reported in great detail, and also the fact that I missed him.

During that period the Trust had a stand at the Bath and West Show in Bristol. There was still a risk that the Royal Air Force would re-open the wartime bombing range on the Dumbles, and Peter had left me to handle the correspondence with the CPRE (Council for the Preservation of Rural England) and any other organisation or individual who might weigh in on our side. At the Bath and West Show I spied Air Marshal Lord Tedder in the VIP seats. Making myself known to him on behalf of the Trust, I begged him to take note of our objections to the RAF's proposal, and he promised to look into it. We were spared the bombing range.

Ray and I had a caravan at the Show and enjoyed our three days. As part of our display we had some Mallards on a fenced pond. All went well until the last evening, when we found that it was not at all easy to catch them up. If we had not had an audience it might have been quite funny.

Holiday time came round again, and as this year Catherine was still in London, we decided to go to Corsica. On 10 June we set off with rucksacks by train to Paris, where we went to the opera before leaving next day by bus for Nice. To our horror the coach was full of English tourists. We decided to converse in Serbian and, if approached by any of our fellow travellers, to speak with foreign accents. The driver obviously realised we were English, but since he did not speak it himself he asked us to translate for the other tourists. It was a pleasant enough drive, with stops for meals and a night in Lyons. We spoke Serbian for two whole days and managed to conceal our nationality.

In Nice we had a day and a half in rather grandiose surroundings reminiscent of Torquay, before flying to Bastia where we found ourselves a hotel for the night.

The following day a bus, which back-fired all the way, took us from Bastia to Pino on the other side of the island. It was hot, and our first requirement was to find the sea. My French let me down, for I enquired the way to 'Le Mer'. The man seemed rather surprised. Why did we want to see the Mayor? Having sorted out my genders we found a gorgeous beach and then some accommodation at the Hotel Allard. (Monsieur Allard had been Chef de Résistance during the war). It was very small and for the first time we were faced with a double bed. As I am a restless sleeper Catherine suffered most — but at least it was cheap.

From Pino we walked right up to the tip of the island (the Cap, where the landscape is bare, wild and rugged), and then journeyed slowly down the west coast. It was a blissful holiday, with sun, bathing and good food and wine. And everywhere there was the smell of the *maquis* — dense growth of highly scented bushes. Sometimes we walked, but buses were our main form of transport.

It rained in St Florent and we did not like the place. The hotel smelled of DDT, and it was from St Florent that through a naïve misconception on our part we narrowly escaped being abducted by Corsican bandits. We moved on in a hurry to Ile Rousse, across a desert and then on to Calvi.

Calvi was special not only because it is a beautiful old city but because we stayed with a Russian family called Kerefoff in the

Bishop's Palace in the Citadelle, perched on the rocks high above the sea. We had an introduction to M. Kerefoff, and he and his family treated us like old friends. There was a party in his house which went on till 5 am, and along with the Mayor and the gendarmerie we saw the sun rise. Madame Kerefoff took us out in the car and up into the mountains with her family and we met some nice Americans from a ship that was in for one day. I wrote in my diary that 'M. Kerefoff is the perfect host, so much so that it was a triumph that we ever got away. His Muscat wine was perfect and his wife and children charming'.

After four days we tore ourselves away. The bus-drive to the spectacular gulf of Porto with its bright red rocks was frightening, and was followed by a thunderstorm and an electricity cut in the hotel after our arrival. We did not linger here but the following day walked the 12 kilometres to Piana and back and eventually found a bus which would take us up to Corte in the centre of the island. It was good to have some cool mountain air, and the scent of the *maquis* on the route up through Evisa was heady. People write about this scent and having lived with it, even for two weeks, you never forget it.

By various near-miracles we reached Ajaccio in time to catch our plane back to Nice. One of the long bus rides was enlivened by the locals singing Corsican songs and telling dark stories about bandits in the area. After our journey through the mountains from Corte, Ajaccio was hot and airless. Smart but dirty, it was not our kind of place at all, but we were not there for long.

The coach trip back to Paris from Nice was a repeat of the outward journey — Serbian conversation, English tourists and all; by the time we reached Newhaven we were unable to stop speaking Serbian.

We had booked seats on the boat train from Newhaven to Victoria but had not been sitting in them long when two charming young men arrived with identical reservation tickets. We pretended not to understand — speaking in broken English — and firmly remained in our seats. The two seats opposite turned out to be unclaimed so the young men sat in them. A good old English tea was ordered for four and we conversed in broken English with our fellow passengers and in Serbian with each other. It was becoming embarassing. Suddenly we remembered the name tags on our rucksacks stowed on the rack. What could be more English than Illingworth and Talbot-Ponsonby? Stealthily we managed to turn the labels round. We protested about the young men paying for our tea. We had indeed woven ourselves a tangled web and it was very difficult explaining that we did not want to be escorted anywhere when we arrived in London. They were so nice and so polite. We felt really awful about it.

Back at Slimbridge, things were much as usual. Diana Johnstone and I drove to Southampton to meet *The John Biscoe*, the Falkland Islands Dependency Survey ship, which was bringing some Steamer Ducks back for the Trust. Ironically, we arrived to be greeted by the rare sight of Sheathbills swimming about on Southampton Water: strange-looking white birds, the size of pigeons, which come from the Antarctic. These had been on their way to a zoo and had been inadvertently released. There was much embarrassment about the episode, but luckily our Steamer Ducks were in good order.

John Yealland, who had helped Peter to install the collection of birds at the New Grounds in early 1947, was now back as curator and was living in Mr and Mrs Bowditch's house across the lane from the cottage. Mr Bowditch worked for the local farmer, Mr Fisher. John had a bedroom over there but had all his meals in the cottage with us. Mrs Bowditch, who was never seen without a woolly hat on her head, had only once been off the New Grounds and that had been when she went to Gloucester by canal boat.

Peter was still away, Douglas went on holiday, and the Trust was running quite smoothly. We were a happy little community. I engaged an elderly membership secretary called Mr Scholes who lived in Frampton-on-Severn. He objected to being interviewed by me but nonetheless took the job and turned out to be marvellous ('with an excellent leg for a boot,' Ray said). He stayed with the Trust until he died many years later.

It was early August before Peter returned, and I spent some time chasing round the country trying to meet his aeroplane. I had just arrived back from one such sortie when I heard the telephone ringing, and Mrs Cameron came rushing out to say: 'It's Mr Scott on the telephone for you'. He had landed at Prestwick and would soon be home. Suddenly it was rather exciting. He had been away for three months, and perhaps I was beginning to realise just how much I had missed him.

That autumn two new characters appeared at Slimbridge. Hugh Boyd was engaged as Resident Biologist and lived in the cottage with us. Two more rooms were added, taking in part of the old cow byre. Lord Geoffrey Percy, youngest brother of the Duke of Northumberland, came to stay frequently and became so interested in the development of the rocket-netting technique that he was nicknamed 'the Pyrotechnic Adviser'. To begin with we used 'Schermuly rockets for saving life at sea' but later we were helped by the army and Geoffrey was sent as our representative to the ordnance factory at Shrivenham where Colonel Tumber made the rockets and supplied the explosives for us.

Peter had visitors every weekend. It was a good life, the Trust was slowly growing, and it was lovely for me having so many friends around, especially Tommy and Diana Johnstone. Diana and I rather fancied ourselves at telepathy and on occasions were rather unnaturally successful at guessing each other's cards picked at random from a pack. I took my elderly Aunt Brucie to dinner at their cottage and after dinner indulged in some palmistry. Aunt Brucie was full of fun, and was thoroughly enjoying herself. I looked at her hand and became convinced that she was going to die before long. I do not remember what I told her. Soon afterwards she did die, and I gave up reading people's hands.

John Yealland was a delightful person and very good company in the cottage. Infinitely polite, he was also shy, and avoided visitors when he could, preferring to be with the birds which he was so interested in rearing. He had a deliciously dry sense of humour and a slow manner of talking which laid itself open to mimicry. He and I spent many evenings in the old studio listening to the Promenade Concerts. John's favourite symphony was Schubert's Ninth — The Great — and I cannot hear it without thinking of him. Music is as evocative of people and places as are smells.

In early 1950 John spent several months in Hawaii advising the U.S. Fish & Wildlife Service on methods of rearing Ne-nes in captivity. His account of his project and how he brought back two birds for the Trust is recorded in the *Third Annual Report* (1949-50) of the Severn Wildfowl Trust. He remained curator until 1951, when he left to become Curator of Birds at London Zoo, and was succeeded at Slimbridge by Tommy Johnstone.

Also recorded in the *Third Annual Report* are visits by HRH The Princess Elizabeth and HH The Princess Marie Louise, who was a very old lady. What is not recorded is that on the damp, foggy day of her visit, Her Highness slipped in the mud and fell in the Orchard Pen. We

Furling the rocket-net with Suzy Thompson-Coon outside a hotel in Scotland; Hugh Boyd in attendance.

were probably more shaken than she was — Peter was away and Maurice Berkeley, our agent, Tommy Johnstone, and I were doing the honours.

The Princess Elizabeth's visit did not go entirely smoothly, either, for this was the occasion when an unknown lady in blue walked along the sea wall from Frampton hoping to catch a glimpse of the Princess. She frightened all the geese which retired to the mud flats for the rest of the day. The Press made a great feature of the happening. The poor 'lady in blue' wrote afterwards saying that she did not mean to spoil anyone's day and she was very sorry. In fact the Princess and the Royal party enjoyed their day which included a splendid lunch for twelve laid on by Mrs Cameron in the kitchen of the cottage. As we came back into the studio the Princess looked out of the window on to the yard and said 'What on earth are all those policemen doing here?'.

Prince Philip had previously visited the Trust with James Justice and had also lunched in the kitchen, where he found that he was expected to carve the chicken; Peter never did the carving, and it was hardly my place to do so. Later John Yealland taught me how to carve birds, and I was often to be grateful for his tuition.

Our first serious goose-netting trip took place in March 1950, for one week only, in Scotland. The team consisted of just five people — Peter, Geoffrey Percy, Hugh Boyd, Julian Taylor and me. Four catches were made, all very small compared to our later efforts.

It must have been on this occasion that I was responsible for firing the net on Lantonside on the Solway and catching only one goose (or was it two — or even three?). Peter was over on the other side of the Nith, and Hugh and I had been left sitting over the net. I was in the dog house. The final catches for that trip were 7 Pinkfeet, 3 Greylags, 25 Greylags and 1 Greylag.

No sooner back from Scotland than we were preparing for a lecture tour in the narrow boat *Beatrice*. Ten lectures were planned for a four-week period beginning the first week in April. The crew was to be made up of friends — varying from six to nine in number. The hard core consisted of Ray, Geoffrey Percy, Peter and me. I undertook to prepare the breakfasts while Ray cooked the main meals. Robert Aickman, Founder and Chairman of the Inland Waterways Association, was to be on board for the whole trip, which he had in fact planned. His knowledge and experience on the canals was essential to the enterprise. Peter had brought a lovely 16mm film of his Perry River expedition and I was to work the projector and sell Trust literature.

Beatrice left Slimbridge on 1 April, but immediately broke down with engine trouble. As two days had been lost, we were always behind schedule and had to hire taxis to take us to the lecture sites. Peter joined the boat at Stoke-on-Trent.

Lectures were given at Macclesfield, Manchester (two), Southport, Liverpool, Northwich, Chester, Dudley, Birmingham and Worcester, and £484 was raised for the Trust. The whole journey occupied thirty-one days.

It was a big adventure with some especially exciting interludes. *Beatrice* was one of the earliest narrow boats to be converted for cruising, with a superstructure which provided living accommodation. As a standard working boat for use on the narrow canals, she was 70ft long and 7ft wide. In her original state she had been an open boat for commercial use with engine and cabin accommodation just forward of the tiller area in the stern.

Working the locks was an enjoyable diversion, but on one particular day a flight of twenty-eight, one above the other, was very hard going. The locks on the narrow canals allow very little margin either side; steering the boat into a lock, therefore, demands considerable skill. Because of the conversion, *Beatrice*'s superstructure continued for the length of the boat, with a small deck area in the bows. Steering her in a crosswind was not easy, and many a time we found ourselves with the nose into the bank, so that much pushing and punting were required.

Beatrice was very beautiful. She had the traditional roses and castles on the panels of the engine room and on the double doors of the main cabin. The superstructure was blue with red and white trimmings.

To anyone who has not cruised the canals it is a whole new way of life. The canals are often higher than the countryside around them, providing wonderful views, and in 1950 there were not many 'pleasure craft' around. Other canal boatmen — mostly working owners — were friends. The greeting was always: 'How do you do?', not 'good morning' or 'good evening'.

Catering was not easy. As the canals wandered through truly rural areas, there were pubs to hand but not always shops. The pubs provided diversion and good company with the locals in the evenings. Returning at night and 'walking the plank' was hazardous — and the

Above: One of my more successful early portraits of Peter, taken on board Beatrice *during the lecture tour, 1950.*
Right: All aboard Beatrice*: Ray Aickman is leaning out of the window, with Geoffrey Percy and me behind her and Robert Aickman sitting above.*
Below right: Beatrice *starting on the passage from Birmingham to the New Grounds to become a floating hostel for the Trust.*

SPENCER.ABBOTT
BOAT BUILDERS
I. C. I.

canals claimed a few soaking victims. Hauling them out in the dark was often hilarious.

Beatrice was the first converted narrow boat to cross the River Mersey. The fifteen-mile crossing between Liverpool docks and Weston Mersey dock seemed a very long way — with such exciting hazards as 'The Seldom Seen Rocks' and the possibility of the wind getting up. Fortunately it was relatively calm and the crossing was made without mishap.

We spent the night at Runcorn near the famous transporter bridge which took cars across the Mersey. From the high towers we had a spectacular view, and Peter went clambering about on the girders — showing that his head for heights was as good as ever, unlike mine. We were taken up in a cage-lift, but I was quite unable to step on to the walkway.

Our worst problem was Harecastle Tunnel. It is one and three quarters of a mile long, and much of the tow path is under water. *Beatrice* stuck fast about 1,100 yards from the entrance. At this particular point the tunnel roof had subsided due to coal workings. The boat was jammed against the roof on the starboard side and the tow path on the port side.

Peter was steering, and we had a full complement on board. He revved the engine and reversed. Then tried again. No go. At the entrance to the tunnel was a large pile of dirty bricks. Wading to and fro along the tow path we carried bricks and stacked them in the main cabin. It was very dark, very dirty, and very claustrophobic. One of our young crew members started to become hysterical and someone had to steady him. At last the boat moved forward — but not far before she became stuck again. This was even worse. Now we could be cut off in front and behind. But at last we drove her through.

Six and a half hours after we entered the tunnel we came out on the other side to find that it was snowing. It was late April. We moored *Beatrice* at the foot of the high embankment, climbed the steps, and invaded the pub at the top. They kindly allowed us to have baths and to clean ourselves up: we were black from head to foot.

The incident made a deep impression on me, and although I had not panicked during the operation — after all there was plenty to do — I aroused the whole boat the following night by shouting in my sleep 'Let me out, let me out'!

Previously, *Beatrice* had been lowered from the Trent and Mersey canal into the River Weaver Navigation on the famous Anderton Lift, a caisson filled with 200 tons of water which can raise or lower boats to a height of 80 feet.

Birmingham has been called the 'Venice of the Midlands'. I am not sure about that, but it is certainly a maze of waterways and looks very different from a boat. You do not see any people, except on bridges, and I think we only once had tomatoes thrown at us.

The last adventure of our canal tour was left to Geoffrey Percy, Mrs Flanagan (who subsequently was to manage the boat as a hostel) and me. We had been delegated to bring *Beatrice* home after she had had another refit. We were on the Severn, nearly as far down as Gloucester. Mrs Flanagan had just passed up some bacon-and-egg sandwiches to Geoffrey and me. We were spinning along downstream, when suddenly a fork appeared in the river ahead. On the bank was a notice with warning of a weir. Too late we took the wrong turning, saw the weir, and throttled the engine into full astern. Nothing happened. The reverse gear was weak. We just went sailing onwards. Now the only thing to do was to go full ahead turning, and make for the bank. For a short time we were broadside on to the current — all 70 feet of the boat, with three anxious figures and the tiller hard over. We could see the lip of the weir very close. But the nose came round and Geoffrey leapt ashore as we hit the grassy edge. The current was less powerful near the shore and we were able to chug back to the fork and proceed on our way to Gloucester, where we locked into the Sharpness canal.

An Ornithologist at Last

By mid-1950 my involvement with the Severn Wildfowl Trust, (or SWT) was increasing. As a budding amateur ornithologist I was able to persuade Peter to take me to the International Ornithological Congress in Uppsala in Sweden in June. It was all very 'proper'. Peter stayed in the University and I stayed with a Swedish family in town. I cannot say I enjoyed the conference, and I can find no record of it in my diary or in letters, but it was a foretaste of others that were to come. The exciting and interesting part for me was the excursions. The first was to Jämtland where we stayed at Handöl on Lake Ann. Until then, Scotland was the furthest north I had ever been. This was beautiful country, with its birch trees, marshes, lakes and hills, and I was thrilled by the first sight of new birds — Longtail Ducks, Scoters, Phalaropes, Divers and Arctic Terns. We went out in a boat and saw ducks on their nests, heard Whimbrel calling, and squelched through a marshy bog where we came upon a pair of cranes. The bog was the quaking variety and we had a slight problem with an elderly, stout, female bird-watcher who started to sink in. With some difficulty Peter pulled her out, while she remained remarkably cheerful throughout.

One day we set out to see Snow Buntings. Our group of about fifteen people was led by David Lack and we were to climb a hill 1488 metres high called Snasahögarna. It was a long climb, with not a bird to be seen. I found myself walking with a delightful but garrulous lady. David suddenly turned round and said 'How can we expect to see any birds if people talk all the time'. He sounded very angry, and I felt very crushed. From then on and for many years I was terrified of David Lack, who occasionally visited Slimbridge.

There were patches of snow near the top of Snasahögarna. It had an inviting peaky summit, so of course I had to go to the top with a small party which included Sir Landsborough Thompson, a mountaineer of some distinction. Peter said that in principle he did not believe in going to the top of mountains so he carefully walked round the top. Once on the summit the usual sense of elation filled us all.

But it was Peter who had the bird-watching scoop: he found the nest of a Rough Legged Buzzard with young.

Two days later, after a long day out from 7.15 am until 6.30 pm looking at Longtailed Ducks, Scaup and Reindeer, we took part in the hotel's Midsummer Day celebrations, with maypole dancing and all the locals dressed in national costume.

From Handöl by train to Abisko in the far north was to be the most exciting part of the whole trip. Peter had been in the area years before and had corresponded with Knut Larsson, the station master at Vassijaure, which is the border station on the railway line to Norway. The 'birding' was good round Abisko and for two days we remained with the group — which by now had dwindled to only about twelve people.

There was a Blue-throat's nest, a Dipper's nest at a waterfall, and the thrill of seeing that beautiful bird the Long-tailed Skua, which dive-bombed us near its nest. On the second day Peter and I broke away from the rest of the party to go further afield. It was a long walk but it produced a wonderful variety of birds, some of them new to me. We stopped at Björkliden and had tea at the hotel. As it was still a long way back over the hill to Abisko, we decided to take a short cut through a tunnel on the single-track railway line. The tunnel must have been about a mile long and it was dark and frightening. So few trains used this line that we thought the chance of being caught in the tunnel was minimal. We were about two thirds of the way through, when we heard a train. For me this was just about the ultimate nightmare. Peter, too, was frightened but somehow he managed to impart a total calmness, and as we flattened ourselves with faces to the wall I felt a great strength coming from him. We were both immensely relieved to get out into the open air again. That day we had covered 30 kilometres.

Next day we took the train to Vassijaure, where we met Knut Larsson and his brother Axel. We were hoping to find and also to catch some Lesser White-fronted Geese and to bring them back to Slimbridge to introduce fresh blood for our captive birds. Knut and Axel owned various boats on the many lakes behind the railway line and also a small wooden hut on Pajep Nurojaure. For two days we scoured the hills and lakes looking for geese, sometimes on foot, sometimes by boat, with an outboard motor, and sometimes just rowing. In all we saw 50 geese, some in flight and some moulted. Those two days were just a preliminary survey, as there were other

Left: Bird-watching in a marsh in Jämtland, Sweden, 1950.
Below: Peter, who had recently acquired a 16mm movie camera, filming a Purple Sandpiper on its nest on Snasahögarna.
Right: By the Dead Lake in the far north of Sweden where we found Lesser White-fronted Geese.
Below right: An attempt to catch Barrow's Goldeneyes on a small, deep pool near Myvatn.

people on the tour who were interested in seeing the Whitefronts. So we returned to Abisko on the second night and made arrangements for the Larsson brothers to act as guides for one or two others in the group.

The next day our companions were Dr Gudmundsson from Iceland and Dr Mills from England. Back we went to Vassijaure and retraced our route by foot and boat and foot and boat to the lonely ice lakes in the hills beyond the tree line. It rained a good deal. Finnur Gudmundsson was very good company and was nonplussed by Dr Mills whom he found rather serious. On the second morning Dr Mills was up at 5 am and Finnur said: 'Dr Mills is very unquiet'. Later, in the same slow, amused way Finnur remarked: 'Dr Mills is always running'.

There were some problems with outboard motors, and the rain and mist were tedious, especially as we had little in the way of spare clothes with us and no way of drying what we did have. The little hut had a primus stove and two wooden benches, and that was it.

My diary records that we 'left Dr Mills on the far shore' and that Axel went on to Inkan. Poor Dr Mills. He did see Lesser White-fronted Geese, and many other Arctic birds, but his visit was too short and he seemed so agitated all the time. The next day Finnur left, but not before we had made plans to visit the Thjorsaver in Iceland to catch and ring Pinkfooted Geese. On the third day, Peter caught our first goose. It was a female, so we ringed and released her, since we only wanted males. Later the sunshine turned to rain and mist, and Knut decided to stay with us in the little hut — which was disconcerting as he snored all night.

On the fourth day we made a little excursion on foot into Norway — more to be able to say that we had been there than anything else. The terrain at the north-west end of the lake was very like a moonscape. It was quite flat, with only low hills in the distance. On the way back we caught three flightless geese on the ice of one of the small lakes. One was a female, so we ringed and released her, but the others were males, so Knut took them to his bigger hut in Inkan, where he made a crate for them. That night it was cold and our clothes were wet. Luckily, next day the sun shone and I was able to take photographs. At that time I had a Rollieflex and no telephoto lens of any kind.

We caught four more geese, of which we kept one male to make up the three that we needed, and ringed and released the others. Then it was more or less a case of 'mission accomplished'. We went back to Inkan, where there were some comfortable bunks, had a large meal, and slept until noon the next day.

We brought the three Lesser Whitefronts back to Slimbridge. They were, no doubt, the ancestors of the birds which we are now breeding and returning to Sweden for re-introduction.

In August the Inland Waterways Association, of which Peter was Vice-President, held a big Festival and Rally of Boats at Market Harborough. Robert Aickman and his staff had organised the week's events. The Severn Wildfowl Trust Narrow Boat *Beatrice* was on show, having been taken up by Geoffrey Percy, and we were able to stay on board. It was not considered appropriate for *Beatrice* to compete for the various cups and prizes, but it was evident that she compared favourably with the other narrow boats and was much admired.

Many distinguished people attended the rally, including Hugh Casson, A.P. Herbert and Gerald Barry (Director of the Festival of Britain). Aerial photographs were required for posterity, so Peter hired a small aircraft and I, armed with a Leica on loan from a friend, had my first surprisingly successful experience of taking pictures from the air. Tight turns are not so disturbing if there is a job to do like looking through a view finder and concentrating on photography.

Among all the activities at the rally Peter took the lead part in a play, *Springtime with Henry*, which was largely made up of a London cast including Barry Morse and Carla Lehmann. His performance was outstanding. Another career which he might have taken up?

The autumn of 1950 was occupied with lecture engagements and goose netting. During that year Peter gave 31 lectures, mostly with the Perry River film which he had made in Canada in 1949. (The original was finally stolen from his car — proving that one should always have a copy made for use at lectures.) We took our own projector with us, which I operated.

The goose-netting forays were now more organised. The object of the exercise was to calculate the total population of Pinkfeet by the method of capture and recapture. Various friends came to help, and the technique was greatly improved. In order to monitor the movements of birds within the country, we dyed the Pinkfeet. The most successful dye was Rhodamine though it cannot be said to have been ideal. The method employed was to dunk the tail-end of the goose in a bucket of the solution, so that the tail coverts were dyed pink. Care was taken to keep the wings dry, and the birds were put in portable hessian cages for fifteen minutes to allow the dye to dry. Geoffrey Percy brought his pale yellow Labrador with him on several trips and invariably managed to end up with not only his own blonde hair dyed but also that of his dog.

Finding hotels to accommodate a team of some ten people, one of whom had pink hair (unusual at that time), whose vehicles were bespattered with mud, and who arrived with filthy jackets and gum boots — probably smelling strongly of goose droppings — was not easy. Arriving after dark as we so often did, we needed hot baths and, for our obliging friends as well as for ourselves, some kind of creature comforts. Our friends were very long suffering. Some came year after year, some came once and not again. Those who were regulars were often known for what they kept in the boot of their cars: in some cases a bottle of cherry brandy, in others a rich fruit cake.

It was hard work. Up at 4.30 or 5 am and a drive to the catching field. Lay the net in the dark, and camouflage it. Set out the stuffed decoys (always Peter's job) and then a long, cold wait. If we were lucky, we would catch before breakfast and go back to the hotel for 'Harryhotters'. More often it was a case of waiting all day, or 'upsticks' and, after searching for the geese, perhaps laying the net in another field in the dark when the geese had gone out. Picnic breakfasts and lunches were normal most days. And it was cold and sometimes wet.

After a catch in a stubble field the nets were full of straw. As many people as possible were deputed to clean the two 60 × 20-yard nets and roll them back on drums: a very dirty and smelly job. If you were lucky, you were invited by the Director (known good-humouredly on these trips as 'the Dictator') to accompany him in looking for the next site for laying the net, or sent out in another search party to follow skeins of geese. In spite of the cold, the lack of hot meals, and the net-cleaning chores, we all enjoyed it. Two weeks at a time were perhaps enough. Longer trips left people tired and scratchy. It was the excitement and anticipation that kept us going. These goose-netting activities went on all through the early months of 1951. During the 1950/51 season 643 Pinkfeet were ringed.

Among our more interesting visitors at Slimbridge was Konrad Lorenz, the famous Austrian ethologist. The first time, he came alone to stay in the cottage, and entertained us endlessly at mealtimes when we were all gathered in the kitchen. Never to be forgotten was the occasion when he stood up in the middle of a meal and scrabbled against a wall to demonstrate how a Carolina duck can climb up a fence. Another year he came with his charming wife, Gretl, and daughter and stayed in the local guest house. When he was not giving a series of lectures at Bristol University he was with us at the New Grounds. He was lively, stimulating and very amusing. At that time he was trying to show that imprinting of goslings on a foster parent was irrevocable. Hugh Boyd was busy trying to prove that this was not so.

Unfortunately there are no letters to help me with details of the period between 1947 and 1951 when I was working for Peter and the Trust. The age of letter-writing was over. Even my diary entries are basic and uninformative. It is difficult now, for example, to bring the freshness of a word-picture to a memorable evening alone with Peter on the moors at Greenlaw, when we watched thousands of Pink-footed Geese pour into the tiny loch. And difficult to remember the small details of events which at the time seemed so poignant or so hilariously funny. The great kaleidoscope of that period was shared with special people like Dougie, Ray, and Diana Johnstone. Most special of all was sharing the excitements of that time with Peter.

It was on 28 March 1951, the day of the Wildfowl Trust's Annual General Meeting and Dinner in London that I knew at last that Peter and I would be married. 'Would be' because the divorce had still to go through, and that would happen in the summer. Strange as it may seem, it was a shock — but what an exhilarating shock and how difficult to keep it secret! Three years had been a long time to wait, and there had been occasions when I had thought of quitting, but I was too involved with the creation of the Wildfowl Trust to tear myself away and venture into the unknown again.

Plans for the expedition to Iceland were advancing fast. James Fisher was invited to join the party. The Royal Society gave us a grant and I was allowed to be included on condition that I did the cooking. It was not exactly my forte, but they did not know that, and I would have agreed to do almost anything to be allowed to go. In any case, there is not much you can do with army rations except heat them, or maybe put some curry powder in the stew.

Peter's divorce case came up in the early summer. He went to

London to appear at the Law Courts. I was at Slimbridge. That evening Hugh Boyd and I sat in the cottage studio after work waiting for the telephone to ring, and I was decidedly nervous. Hugh knew why Peter was in London — but perhaps he did not know the implications for me, though I suspect that he did. The decree nisi was granted. There were no complications, but there would be a period of waiting before the decree absolute came through. Peter sounded relieved, and so was I. Hugh smiled happily and I felt how good it was to have friends around.

James Fisher went to Reykjavik by sea with the supplies and equipment for the expedition, and Peter and I followed by air on 12 June, courtesy of Loftleidir, the Icelandic airline. We stayed two nights with the British Minister. The British Legation was at Höfdi — the lovely house near the shore where Presidents Reagan and Gorbachev were to meet many years later. Peter had known our host, Jack Greenway, before. Jack was unhappy in Iceland having previously been British Ambassador in Panama. He preferred hot climates. That first night at dinner he talked of retiring, but said he had only one regret: he was authorised to conduct marriage ceremonies, but had never had the opportunity. Peter and I felt that it was a pity to deprive him of his last ambition, so we asked him if he would marry us. There was, however, still the little matter of Peter's decree absolute. A telegram was sent to Peter's lawyer, Keith Miller-Jones, who was an old friend of the family. I wrote to my mother saying that I planned to get married but that it was a secret. I wrote to Mrs Cameron at the New Grounds asking her to send out various clothes, including a white broderie anglaise dress which I had bought while window-shopping with Ray not long before. I did not explain why. Then off we went to Myvatn, where again I heard the magic call of Longtailed Ducks. The weather was fine and the lake was staggeringly beautiful. The water was covered with ducks and many other birds, and the shore was lined with kingcups and other wild flowers. We spent four nights there and then returned to Reykjavik via Akureyri, where we flew out in a Catalina flying boat from the nearby fjord.

Meanwhile Keith Miller-Jones had replied that the decree absolute should have come through by the time we were back from the expedition, and that he would send it by airmail to Jack Greenway. Jack said he would try to organise an air drop into the camp on the Thjorsaver with letters and food, so that we would know.

Setting a net to catch Harlequin Ducks on a fast-flowing river at Myvatn in Iceland.

Success in Iceland

Our expedition party consisted of just the four of us — Finnur, James Fisher, Peter and me, with one Icelandic guide who would remain with us to help with the ponies after our arrival in the Thjorsaver. Finnur was a giant of a man. He must have been 7 feet tall and correspondingly broad, with short, greying ginger hair, a neat beard, blue eyes and a quiet, delightful sense of humour. At that time he was 'Mr Conservation' in Iceland and an excellent all-round naturalist.

Peter had known James Fisher for many years. James's father was headmaster of Oundle when Peter was at school there. I had met James several times but knew him only a little. He accepted my presence in the party without question, and played a full part in the organisation and arrangements. He was a wonderful companion, with a robust sense of humour — perhaps rather too fond of puns, but then Peter could join him in that. He was also extremely considerate and always willing to help with any problem, however small. Occasionally he even offered to cook and wash up.

The story of the expedition has been told in *A Thousand Geese*, written by Peter and James. My version of events is perhaps a little less academic.

We set off on the expedition on 22 June after an excellent farewell lunch at Höfdi, having finished our packing in a great state of excitement. Finnur Gudmundsson collected us from the Legation and took us to the Museum, where he worked and where the bus was to collect us. It was 6.30 pm before we finally departed. We filled the thirty-two seater bus: just the four of us, with our army rations and equipment. The road, like all roads in Iceland, was just a dirt one and very bumpy in patches. Finnur was still concerned because we only had one kettle and no teapot, and he insisted on stopping in a small town to try to buy one — but the shops were shut.

It was an interesting drive, through lava fields, marshes, and farming country, and past a town whose water — including that of many acres of greenhouses — was heated by natural springs. As

Left: Finnur Gudmundsson, our beloved expedition leader. Here he is recording the ring numbers of the geese as they are caught.
Below left: James Fisher, the intrepid horseman, cooperatively posing for me. The pony is standing still for once.
Below right: Peter posing on a mound on the tundra.

Finnur said, there was no cold water — and indeed clouds of steam rose from various points all round the town.

We arrived at our destination at 9 pm: a lovely farm house called Åssolfstadir where there were hordes of children, but one portion of the house was reserved for guests.

The following morning we re-packed *all* our luggage in a barn. It transpired that we had far too much for the seventeen available ponies to carry, so we had divided it into three parts — 'essential', 'second trip', and 'leave behind altogether'. We also had to empty out our rucksacks, as they were not suitable for strapping on pack ponies and were too heavy for us to carry on our backs while riding. Everything was stowed away, then one of the guides pointed out that two of the kitbags were no good because they rattled when shaken and would scare the ponies. So they had to be re-packed: all the billies wrapped in paper with their lids tied firmly to the kettle. By this time it was 12.30 and lunch was ready. The guides said it was too late to ride to our first camp so we should hire a lorry to take us there. Time is fairly meaningless in Iceland and the ponies didn't leave until about 4 pm. We lay in the sun in front of the farm and watched them go down the hill, on to the dried-up river bed, up again, and out of sight. Mount Hekla was looking at its best in the sunshine after a day of rain. Plumes of smoke rose from the craters on the west side and clouds rolled intermittently over its summit. It seemed too good to be true that the weather should be so bright. It had rained all the way from Reykjavik to Åssolfstadir.

Half an hour after the ponies had left, a cry went up and the farm truck went hurtling down the hill. Apparently some of our ponies had broken loose and had made for home; but eventually we were reassured that all was well. No-one seemed to know at what time the lorry was supposed to be coming for us, but at 6 pm a telephone message came to say it had broken down. As it was the only vehicle in the neighbourhood large enough to carry us and our equipment, the situation seemed a little critical. However, the farmer said that he would take us by Jeep and tractor. So one trailer was loaded for the tractor and another for the Jeep, and the four of us, with the driver and our most precious pieces of luggage, were squeezed into the Jeep.

There was no real road — just a track — and we had only been going about an hour when, in the middle of a vast, barren lava field we had a puncture. There was one spare wheel, which was quickly changed. We caught up with the ponies and arrived at the camping site about 9 pm, but not before the track had become so bad that we all had to get out of the Jeep. The men pushed while I held the seventeen ponies.

We unpacked our tents and put them up, including the big mess tent. The cooking utensils also had to be unpacked and I made a meal for the six of us from our army ration pack consisting of bully beef, potato powder, mixed diced veg, and fruit. Next day the weather seemed to have changed for the better and we awoke early with the sun blazing on to the side of our tent and a very great heat inside it. It was all rather difficult because between midnight and three it was very cold and I woke up every two hours.

After a cold breakfast we packed up again and by 10 were ready. But it was another two hours before we left. The packs had to be made up by our two guides and strapped to the ponies. The ponies were then tied head to tail, and off we went. I was allotted a nice little grey. Finnur had the skewbald — the only 'large' pony at about 14 hands. The others could not have been more than about 13.2. James had one which was rather fresh and wanted to be in front all the time.

The sun shone and the wind blew so that the dust came in clouds behind the ponies. We were supposed to ride behind each line of pack ponies to keep an eye on them and the packs. We had already seen terrific sandstorms whirling past us at the camp along the stretch of desert land near the river Thjorsà. Although we didn't actually get caught right in one, it was bad enough. However, it was all strange, new and exciting. Of course I was delighted to be on a horse. My beast was quiet enough, but full of go. They walk very fast for their size and take such short steps that one cannot rise to the trot: it is in fact a 'triple'!

At Camp 1 we saw and heard quite a number of geese, which was encouraging. We stopped for our second camp at about 6.30 pm, having only eaten a few biscuits, just off the track down a hill towards the Thjorsà — a lovely place. The ponies were unloaded, hobbled, and untethered. There was a mixture of low willow (about 9 inches high) and grass on the oases for them to feed on. There followed the same procedure as before — with packs, tents, food and repacking. The guides declared that they didn't like our food — especially the crispbread, so we opened up the wooden box containing the biscuits.

Once again we were ready by 10 am next day. But this time, they had to get the heavily loaded ponies up the steep hill behind the camp

Left: When it was hot and still, the flies were a menace, but would always buzz around one's highest extremity!
Below: Pony-train crossing the desert on the way to the Thjorsaver.
Right: Peter writing up his notes, and James resting, at Camp 1 on the way to the interior.
Below right: In the mess tent at Bolstadur: Peter shaving, James reading, Finnur plant-pressing.

on to the route again, so six animals were led up while some of us kept an eye on the rest. The guides came back to pack them, and while this was happening one of the ponies went wild and bolted over the hillside, casting the pack and pack-saddle all over the place and breaking its three girths in the process. As they had to improvise new straps and girths, this set us back by half an hour. At the camp we saw a goose (a Pinkfoot) sitting on her nest on the far side of the river Thjorsà, which caused quite a lot of excitement. Finally we started off — riding across rivers, desert, hills, valleys and oases for another seven hours. We were pretty stiff, and James had fallen off the day before because he had tried to read his map while mounted. The ponies wouldn't stand for anything unusual going on, and his just went mad. James didn't hurt himself, and the pony was easily caught, so all was well.

On the way to Camp 3 Peter, Finnur and I stopped to look at some Whooper Swans. They were nesting and we wanted some photographs. As this took us about half an hour or more, the pony convoy was miles ahead when we set off again. We rode for some time and then lost the track on the vegetation near a river. It was a bit worrying, and we were looking desperately for tracks when Peter said that his pony kept pulling to the right: he was sure that the convoy must be over there.

Sure enough, a few minutes later we saw the ponies stationary on the far side of the hill. Johann rode to meet us and showed us where to cross the stream. To try and satisfy the guides, who were craving for meat, we opened six tins of stew. I don't think they really appreciated it: but we did. After a small (though cooked) breakfast and only three or four biscuits for lunch we were pretty hungry by supper time. The guides had the exacting task of watching the ponies all night to make sure that they didn't set off for home on their own. That night we saw and heard quite a number of geese, both Pinkfeet and Greylag.

We decided that we must have a long lie-in, so we slept late and didn't leave until 3.30 pm. It was an exhausting day. For four hours we rode over real desert. It was very hot and there were mirages everywhere. James's horse lay down with him in the lava sand and seemed to have given up the ghost. So James walked for half an hour, and after that the pony carried him again. Our faces and hands were burned but, luckily, for the most part we rode with our backs to the sun.

At about 7.30 pm we stopped at a small oasis in the desert and unloaded the ponies. The flies were so bad and the ponies so troublesome that we loaded up again and went on without food. After that, more desert — and Hekla now looked very far off down the valley of the Thjorsà. We waded through rivers — of which one was especially wide and deep and was reputed to have quicksands. Before we set out we had been warned about the likelihood of our ponies going into quicksand and had been told to follow the guides and pony train exactly.

Further on there was once again some vegetation, but patchy and scant. On we rode, mile after mile. We found a Pinkfoot's nest with babies, and photographed them. Still we rode on, now right along the edge of the Thjorsà, until finally we came to the edge of the big oasis which starts at the foot of the moraine south of the Hofsjökull. Here we stopped, right in the middle of a breeding colony of Pinkfeet and 150 yards from one of the three big tributaries of the Thjorsà which divide this big oasis. I was too exhausted from lack of food to care about anything. The difficulty was that the guides refused to take us further. Peter was, of course, so thrilled with the geese that he was beyond caring about food. So as soon as the ponies were unloaded I sat down beside our untouched lunch pack and ate and ate.

The wonderful fine hot weather had caused the glaciers to melt, so the rivers were high. All the rivers in this area had quicksands — bad enough for a human, but even more of a menace for a pack horse. Johann tried our river (the Blautakvisl) and declared that it was impassable. Finnur said that the guides were tired and bad-tempered and that we had better not try to push them, so we let them sleep. We put up the big tent only, and Peter, James and I slept in a row while Finnur watched the ponies.

The idea had been to give the guides a rest and to take shifts watching the ponies through the night, but Finnur said that he hadn't the heart to wake Peter or James, and he watched for the whole night himself. At 6 am another discussion with the guides took place, with Finnur interpreting. The outcome was that they were to ride higher up the river and to try to cross the Blautakvisl while we watched the ponies. So we had breakfast and they then set off. As they went over the hill we heard a cry from Finnur and saw one of the hobbled ponies hurrying down the track along which we had come. We were still eating breakfast but went to see what was happening, though we thought Finnur was panicking a bit. Another pony followed, and in two minutes they were both proceeding at this agonizing hobbled gallop along the track. The men ran after them, but it was no good:

Peter with a Pinkfooted Goose which has just been ringed.

so off they went to round up the remaining eleven ponies (the guides had taken four). They caught them all, and James and I carried a saddle over to them. I then set off on my grey after the two runaways. They had a quarter of an hour's start, and by the time I reached the top of a hill near the track they were nowhere in sight. I rode as far as the stream where the Pinkfeet babies had been photographed the day before, and saw the pony tracks in the sand. It was obviously no good. Finnur had said that they would gallop all the way home — and I'm sure they did. Peter came after me on the skewbald and we rounded up a black pony out of a marsh and drove it towards the camp. We couldn't catch it, and it had no halter.

Peter and I then set up a hide to photograph a goose on her nest 100 yards away from the camp, while James had a bath in a little iron stream (bright yellow water) and watched the remaining ponies in a sheep kraal. I returned, had a bath and washed my hair in the same stream — where, in a semi-clothed state, I was startled by Johann returning from up river.

More discussions took place. It appeared that the guides had crossed the river higher up, but said that it was too deep. It was another five or six hours' ride to the place which we originally intended to reach, and if the rivers were bad it would not be worth risking the pack ponies with all our precious film equipment, radio set, and clothes. So we agreed to stay at the camp (Camp 4). We had a meal and set up the tents, then the guides departed, taking all the ponies. So there we were — three days' pony ride from civilization. The sun blazed down and I sunbathed. Our faces were bright red from the three days' ride in the sun and wind.

An amusing diversion when dealing with Iceland ponies is that on dismounting you take the reins over the pony's head and drop them. The pony then walks about or stands, but watches the reins the whole time and never treads on them. They can even walk quite fast with their heads out sideways.

We spent the next week exploring the neighbouring marshes. There were a lot of geese — all Pinkfeet. The first few days some still had eggs, but later we saw more and more with downy babies. We marked some 33 chicks with small aluminium tags: they were enchanting little creatures — greeny yellow and grey, just the colour of their background. We also found quite a number of Whooper Swan nests, huge great mounds with enormous eggs.

The first night there was bitter disappointment when the radio failed to work. The second night, however, the receiver was made to operate and we could hear the Post Office at Reykjavik calling 'Hofsjökull Expedition'. The next morning Finnur took the whole box to pieces and put it together again. In the evening James had another go, and on the third night we established communication with our radio contact. We told them to inform the British Minister that we were now in radio touch, so that he could pass messages if necessary. It was very satisfying to have the equipment working, and we felt much elated. On the Sunday after we arrived it rained all day, so we stayed in the big tent, and I darned socks and read and played pegotty with Peter.

One of the most difficult factors to contend with was the sudden changes of temperature. If the sun shone it was hot, and out of the wind, very hot. If the sun went in, it became cold quite quickly. The wind was nearly always cold. Round about midnight it was very cold indeed. I slept in all my clothes, with James's flannel pyjamas over them. The only garment I removed was my trousers, but I then put on long, thick Naval woolly pants. Yet if the sun shone first thing in the morning it became terribly hot in the tent and we would lie in a stupor, gasping for breath.

On Wednesday 4 July we set out on our longest trek, at about 11.45 am, returning to camp at 3 next morning. We wanted to prospect the moraine at the foot of the Hofsjökull to determine if it would be a good idea to move camp there when the ponies came. This walk involved crossing three rivers, all tributaries of the Thjorsà. The first was almost at our doorstep and we crossed it most days. Peter had thigh-boots, so was well set up, but James, Finnur and I wore mackintosh trousers tied round our gum boots. This sometimes worked and sometimes didn't. If the water was above the knee and was very fast-flowing it seeped in.

The quicksands in the rivers varied from day to day and so did the height of the water. After a hot day it would rise as the glaciers melted. All the tributaries on that side of the Thjorsà were of glacier water, which was not really suitable for drinking. It was creamy white and carried a fine sand in it. The second river, two miles further on, was the worst and we were quite a time negotiating it. We crossed at the mouths, where the rivers were widest, but the sand and mud were tricky, and there were quicksands.

The rivers divide the Thjorsaver oasis into meadows or 'ver' and the *ver* which we had now reached was called Illaver (pronounced 'Itlerver') which means 'the bad meadow'. And indeed it was bad: very swampy, with fewer willows growing along its banks. At every step one sank three or four inches into quaking bog. As we crossed the marsh it was very hot and we wondered if we would ever reach the glacier. We found a group of moulting geese, but we couldn't do anything about catching them as they saw us long before we were within running distance of them. After one more river — which, as usual, I was carried across by Peter — we reached the foot of the glacier.

There was an old moraine, just like foot hills, and the vegetation here was rather lush in parts, with lovely patches of flowers, in particular *Sedum rosea*, a succulent. Behind the old moraine was the newer one: great hills of shingle, like the mountains of the moon, with no vegetation and deep pools of water. It was very grim and forbidding, and stretched for at least a quarter of a mile along the edge of the glacier. James and Finnur went up to the glacier while Peter counted the geese on the marsh below, and then we walked round the edge along the old moraine.

At about 9.30 pm we set off for camp. At about 11 pm the sun went down behind the Hofsjökull above the glacier and it became very cold. On our return to camp at 3 am we had a huge meal consisting of sausages and bacon and cake.

Two days later we set off again, this time down the Thjorsà. We followed another tributary and came upon some goslings which we caught and marked. We also marked 3 Whooper Swan cygnets which are quite the most enchanting of all the downy young. They are so soft that you can't feel their bodies through the down: softer than the softest baby's toy. Whoopers are bad parents; these left their young when we were still 150 yards away and didn't stay to protect them as geese would have done.

Our main tent was very homely. The four of us each had a ration box to sit on in our own 'patch', with enough space to stretch out on the floor and lean on our things. The floor was mostly of moss — and bits of dry moss found their way into almost everything — but it was clean and dry. The kitchen occupied one corner of the tent and the radio the other.

The next three days were rather uneventful: with the anticipated arrival of the pack-ponies from the second trip preoccupying all of us.

On the 9th it was decided that we shouldn't go far from camp as the ponies might arrive that day. The weather was showery, and sudden changes in temperature were very marked. We sunbathed — and also sat in the big tent with all our clothes on. These last three days had been very dull in comparison with the early ones when we were still exploring and finding new things every day. It did occur to us that we would be in a pretty mess if something went wrong and the ponies didn't turn up. We had food for six days after the 10th, when they were due, but it would take four days to walk back and by now the radio set was not working at all reliably.

However, at around 7 pm we were sitting in the big tent discussing the situation when the ponies arrived. Relief supplies! Nescafé, jam, rice, dried fruit and curry powder: all of which had been left behind at Åssolfstadir for Trip II. We were now stocked up with food for another thirty days if necessary, and with our ponies and our new guide Valentinus, or Valli as he was usually known, we would be able to cover a much wider area.

A fox hunter turned up with the two guides and twelve ponies. In those parts they shoot as many foxes as possible, because the sheep spend the summer up there unattended. We saw only one Arctic fox — and, of course, with our interest in the geese we were only too glad if the hunter shot the foxes. But he went to all three known earths and saw no trace, so we hoped that on his travels he hadn't disturbed the geese too much.

The arrival of the first sheep, incidentally, was quite an event. They are let loose at the farms and wander off by themselves, eventually reaching the oasis. They are then rounded up in September by the farmers. The two we saw were very wild and didn't like the look of us and our camp.

The day after the ponies came we took them southwest to a tributary called the Hnifa, where we caught quite a few geese, including our first moulted adult, which pleased us very much. The goslings had grown a lot in the past few days and could now wear rings instead of wing clips. The ones we caught were quite big and could run as fast as we could. As soon as they realize that they are definitely being overtaken (and with ponies this is possible) the adults crouch and flatten themselves on the ground. Even then, there is quite an art to catching them. You have to approach quite slowly, pretending not to have noticed them, and only when within a foot or two is it safe to grab them. Otherwise, if you run up to them they rush off again at great speed. Valli was marvellous. He galloped madly after them and caught more than we did. We put them in sacks and then released them all together, so they ran off in a little flock.

As we had two pack-ponies as well as our own, we had to take them with us wherever we went; they couldn't be left alone at camp in case they made off homewards. However, we used one of them to carry our lunch and rain clothes, and Valli led them both.

My pony (a mare) was called Hedja ('Heroine') and was a very good one, better than the grey I had ridden on the way up. She was dark chestnut, stockier than the grey, with a great thick tow-coloured mane. Finnur had the same skewbald, Peter had one the same colour as mine, and James had a bright chestnut with a very pale mane and tail. His was very lazy, and he was always left behind. We were altogether well pleased with the ponies and full of hope for our goose-catching.

We spent the next day in camp, as it was cold, wet and windy. The most exciting event of the day was a radio message from the British Minister saying that he was trying to arrange for a plane to drop mail, etc., on Monday.

The next day the weather was a little better, though still cold and showery. We crossed the Blautakvisl on the ponies and set our nets on a hill.

The netting was very much in the nature of an experiment, and was in fact a failure, though we learned much from it. We tried to drive too big an area, and the net was in a dip on the shoulder of the hill. We learned that when geese see danger approaching they always go to the top of the nearest high hill, so the nets should have been put as near to the top as possible. Some of the party had seen a number of flightless geese on another nearby hill.

So after having a laugh at our failure, we picked up the nets and moved them to the other hill. We set them just over the top, and left them for another day when we planned to drive the geese from the lakes on either side.

The tops of the hills were bare: just stones, sand and shingle, and the geese probably went there for three reasons: (1) they were much more difficult to see on the stones, (2) it was much easier to run there than among the dwarf willow, and (3) they could see all around them.

Above: Finnur, James and Valli (back to camera) outside the mess tent at Bolstadur, saying goodbye to the pack-pony guides.
Right: 'Cedric, who shall be nameless'. The first Whooper cygnet to be marked.

Left: Valli saddling his pony, who was the liveliest of them all. Below: My pony, Hedja. When the reins were dropped she made no attempt to run away.

We had been told that there were various 'goose folds' in this area and that nothing was known about them except that they bore this name. How they worked, no-one knows, but it seemed certain they must have been used for catching geese as far back as the 15th Century or earlier.

It was not until the following day on a hill very near the ice cap that we found our first goose fold. Though one of our guides had marked on a map the spots where they were situated, we had failed to find any before. This one was right on top of a hill, but in a slight hollow — so in the old days they evidently knew about geese running up hills. The 'fold' consisted of a narrow horseshoe of stones — a crumbling wall — about ten yards long and two yards wide. We spent much time wondering how they got the geese to go into these folds. It seems unlikely that in those days they had nets, yet, if not, how did they lure the geese into the narrow catching cages? And if they had nets, why not make the cage of netting too? It was all very mysterious, but intriguing to think of them centuries ago knowing all about the habits and behaviour of geese and doing the same thing as we were for a different reason.

Crossing the rivers on ponies was less laborious than wading through them on foot, but quite a challenge and somewhat alarming. The ponies floundered around in the soft places, and hated going over bogs, streams and rivers.

One of our most rewarding expeditions was to a new 'meadow' near the corner of the moraine. On the way there we carried out our drive for the nets that we had left earlier, which was very exciting. As Peter and I came over the top to view the lake on the side of the hill where the nets were, we saw the geese already up near them. About 50 escaped and crossed the river, but we caught 15 — 7 adults and 8 young. We then went on across the river and caught about 20 by riding after them up another big hill.

As soon as geese were sighted, Valli dropped the pack ponies and galloped off. I usually followed next with Finnur — at the gallop full tilt across the tundra. The young geese usually got left behind and Peter and James were in the rear picking them up as they came.

On our way to the hill Peter and I saw a Snowy Owl. They are not too common and are very strikingly handsome: huge white birds with dark flecks and feathered trousers. Our companions were most envious, and we were lucky enough to see the owl again in a completely different area.

Left: Peter holding two Whooper cygnets while Finnur meditates beside the Thjorsà.
Below: Ancient goose fold near the top of Nautalda.

On the way back from Arnarfellsalder, where we had caught the 20 geese, we met some more on the point where the Blautakvisl runs into the Thjorsà — just below our camp. Luckily it was right by the crossing, and Valli hurtled into the river, catching 2 adults. The others followed and caught some young while I stayed on the far shore and caught 2 going back. I took a sack from Valli, and was coming across the river shouting for instructions about the route to Peter, who was rather more that half way across, when suddenly in midstream my pony just went down. I found myself half submerged, but still clinging to the sack which contained the goslings. Curiously the bottom wasn't soft — the pony must just have struck a soft patch with her left foreleg. She rolled over on her side but righted herself again. Valli was behind me and collected the pony while I splashed to the sandbank where Peter was standing. I was soaked from the waist downwards. My windproof jacket had kept most of my top half dry, as it was pretty waterproof, but my Exakta camera, binoculars and watch had all been immersed. The watch and glasses seemed none the worse, but water had got into the camera.

Most of the unfortunate incidents with ponies seemed to happen to James. One evening when we were riding along to the crossing at the mouth of the Blautakvisl, James's pony went too near the edge. The ground was soft and boggy anyway — pretty treacherous going — and it just fell away where he trod. The pony floundered and finally slipped into the water. James jumped off just in time, and the pony was then seen by all of us swimming down the deepest part of the Blautakvisl towards the Thjorsà. It was quite the funniest sight, and we all roared with laughter. The pony seemed unmoved, and was drinking as it swam. Valli caught it near the beginning of the crossing, and all was well.

On Friday 20 July, after resting the ponies for a day, we packed up our three light tents and enough food for three days and set off for the corner of the moraine. By 'corner' I mean the edge of the terminal moraine — where the moraine was at its widest. On the way to the hill where our nets were set, we came on a large number of geese and made our biggest catch — 28. We went on to organise the nets, but it was not a success: the geese were too far away and went up the wrong hill. However, we did manage to catch seven.

Two of the pack-horses were very heavily laden and had to be watched or held all that day. We had one more goose-chase on a hill, then a final one to catch six goslings on a lake in the vast marsh beyond. While running after one I went right down in the bog and fell flat in the spongy moss, getting soaked to the waist again. I dropped the gosling I was carrying. My other camera went down under me, but didn't seem to get seriously wet. The water was very cold and I felt pretty wet and miserable. We continued for about a quarter of an hour and then camped under the outer moraine. There was a cold wind blowing, but we had the tents up quickly and the primus going in the shelter of some sacks slung between the tents. We were pleased with the good water from a spring in the marsh near by — a nice change from the bog water at the base camp, which was seething with live food. While supper was cooking, James caught one more gosling, bringing our total for the day to 53 — a record.

It rained during the night, but was not too bad when we got up at around 9 o'clock the next morning. We rode round the edge of the moraine, then crossed a large expanse of shingle and a river, and came to the foot of Arnarfell — the highest peak rising out of the glacier. At this point the glacier was indented before jutting out in another great curve, and here at the foot of this mountain we found a remarkably lush expanse of vegetation. It was a riot of flowers — a carpet of bright yellow saxifrage on the flat at the bottom and among the stones and moss; and a mass of geranium on the upward slopes.

That day we had some wonderful views of the White-tailed Eagle which had shadowed us since the morning before, when we found it eating a large gosling. It was a huge bird — and looked rather impressive at Arnarfell, sitting under a rock on the steep slope of the mountain. (Arnarfell means 'Eagle Mountain'.)

After lunch we left Valli to look after the ponies and walked up to a small glacier behind Arnarfell. On the way we found a fascinating ice cave in the side of the hill. For a long distance up the mountain there is ice lying only one foot beneath the scree, and the scree was coming down in small avalanches beside the cave and rattling into the water inside. We watched a river coming out from underneath the ice on the small glacier and saw a moraine being formed by the glacier. Note that the glacier was small only in relation to the vast Hofsjökull.

We were going on along its foot when I discovered that I had lost my watch. So James and I retraced our footsteps while Peter went on with Finnur. It was pretty hopeless looking for such a small object, and we didn't find it. Its value was largely sentimental as I had acquired it twenty-two years before from cigarette coupons given to me by Miss Tin.

When Peter and Finnur returned from their botanical walk we started back for the camp. The weather was still cold and showery. We rode along the top of the outer moraine, counting goose nests. We went at a great pace, as the ponies knew that they were going home and had had an easy day. It was quite exciting cantering up and down the grassy mounds as though on a series of switchbacks. After another dinner of stew cooked in the entrance of my tent, we retired early to bed and were up in rather better time than usual next morning. We struck camp in record time and were away soon after 11 o'clock.

We continued west along the moraine counting goose nests and then crossed the Miklekvisl on to the more marshy Illaver. We had one exciting goose hunt, with the ponies mad keen to gallop, but most of the geese disappeared very mysteriously into the mirage. We caught about 9 goslings and ringed them. By this time a strong, bitterly cold north wind was blowing, so we put on all the clothes we had and went on to Nautagi ('Bull's Meadow'). This was an area about sixty yards square with hot springs and wonderful lush grass. Finnur found some very rare plants and several new ones.

After lunch a little further on, we rode round a hill (Nautalda) and had a terrific goose-hunt over the old river bed which was shingly and soft in patches. Valli and I galloped far ahead over it and caught quite a number of geese.

After that we had one more chase and arrived at base camp around 8 pm. It was pleasant to be 'home'. Although near the edge of the big oasis, we nevertheless seemed to be much more centrally situated for goose operations than we would have been anywhere else. It was a very cold night with quite considerable ground frost — about the coldest we had.

Next day we rested. It was cold and windy and the Thjorsà was very low. Valli, who had never made the crossing before, performed his test-crossing on the bigger of the two pack horses. All went well and we resolved to cross over the following day (24 July).

So the next day we put on our macintosh trousers and set off to cross the Thjorsà just by our camp. Having been down with my pony in the Blautakvisl I was a bit scared — but we all crossed dry. The water came to about the middle of the ponies' tummies and not quite over our gum boots. On the other side we caught 57 geese and as a result didn't get very far. It was rather unfortunate that the geese went up very steep hills, and the ponies were soon exhausted by so much galloping on the stony ground. Worst of all, James sprained his ankle very badly running after a gosling, and nearly fainted with the pain. So we came back slowly. The river had risen several inches, but once again we all crossed safely. Much credit for this crossing was due to Valli. He had said before that he himself wouldn't hesitate to cross it when low, but he wasn't sure about the rest of the party who were, of course, inexperienced in river crossings of this nature. We were given instructions about what to do if the ponies stumbled, sank in deep mud, or started to swim. Apparently they don't swim until the water is over their backs — then you hold on to the mane as well as the reins!

The next day, Finnur, Valli, Peter and I set off for the hill near the Miklekvisl where we had left the nets. We split up, Valli and I going one way and Peter and Finnur the other. The operation was all carefully planned, with a strict time schedule. I had the shortest distance to go and an hour to wait. Two minutes before the appointed time, I saw Valli appear on the far edge of the marsh, so I leapt on my pony and rode across the hillside. Several groups of geese came into view and it was very exciting to see them streaming across the marsh and up our hill. We all closed in slowly until finally we were all round the hill. Then suddenly I saw a vast quantity of geese streaming straight down the hill on the net side. I couldn't see the nets and guessed that they must have blown down. At a signal from Peter, who appeared briefly on top, I galloped round and met Valli who had headed off the geese. In the meantime, Peter was putting up the net which the geese had broken down. Having done so, he withdrew as the flow of geese came up the hill in front of me. Then I saw how they had made their way through. They had all piled up in one place on the net until the ones at the rear could just walk across the backs of the front ones struggling in the net. Once more, the net gave way, and there they all were on the wrong side of the cage again — with the exception of a few who had their heads caught in the net. By this time Finnur and Valli were on top of the hill, so the geese stopped in a tight bunch. Peter and I extricated the trapped ones from the net, ringed some, and put the rest in bags. Then we set up the net again carefully while the geese just stayed in their tight flock on top of the hill. Finnur was in a terrible state of anxiety, as he and Valli were responsible for holding the geese while we were busy. But they held — and then slowly, slowly, the flock was driven into the cage. We weren't sure if they would all get in to the small space, but they just

Left: Valli holding two Pinkfeet while Peter puts on the rings.
Below: James, now bearded, with a catch of Pinkfeet on Arnarfellsalder.
Right: Peter silhouetted against the edge of the Hofsjökull icecap.
Below right: A wind-swept Peter with his rucksack, ready to set off across the tundra.

managed it. Quite a number escaped — probably about 30 all told. The pressure was so great at certain points that we had to keep strengthening weak parts.

I climbed into the cage, caught the birds, and handed them to Valli who held them while Peter ringed them. Finnur wrote down the numbers, sexes and ages. We only had 120 rings so when we saw that there were many more birds and only 10 more rings we sent Valli back to camp for a further supply. In the meantime we withdrew ten yards, and sat down. The remaining geese moved from the edges of the cage, and all the young ones sat down in the middle and went to sleep while the adults stood quietly around.

Valli was back inside the hour with hundreds more rings, and we set to work again. We were very excited to find two geese in the cage with blue rings. These were birds that we had handled in Scotland last winter. The total catch was 267. It was 5.45 pm before we had finished, and we retired down the hill to have our lunch. Afterwards we strengthened the nets again and left them for another catch.

The next day, Thursday 26 July, we went down the river to Eyvafen, and caught 44 geese. The most amusing thing we did was to set a net (Valli set it) while keeping 16 goslings a few yards away in a bunch. We then drove them in. We also set the net on a hill and tried a drive, but it was unknown country and the operation was not very successful: the catch being only 17 when we might have had 100.

On Saturday 28 July we set off for our net hill again. There was still a strong, cold north wind blowing. All went according to plan again, with a slightly different timing schedule. There were, however, fewer geese in the marsh and we caught only 97: though this was very satisfactory. We had started early and finished ringing by 1 o'clock.

We crossed the Miklekvisl and had lunch at the foot of Arnarfellsalder by the big lake which we named 'Vallarvatn' after Valli. Then we climbed up the hill and Peter went to prospect for geese. He came back with the news that there were a lot between the hill and the moraine, so we proceeded cautiously all round the back of the hill to a small hollow near the far end of the top, and set the net. It was only a small net and cage — the total length being 80 yards — so that it had very short 'wings'. Then Valli set off one way round and I the other while Peter and Finnur were to act as flankers. Unfortunately, I found myself ahead of Valli and saw hundreds of geese streaming up the face before he even came into sight. He went further away from the hill to

Finnur dips his thermometer into one of the hot springs at Nautagi, while Peter writes down the temperature.

The aeroplane which brought good news and extra supplies.

round up a group of geese heading for the moraine, while I was left standing — having headed my geese in the right direction — between him and the hill. By now many of the geese must have reached the top and I couldn't attract Valli's attention to make him come back. So there I stood while Valli chased about 30 geese, knowing that at least 300 had already gone up, and only Finnur and Peter were there on the very broad summit to hold them. Eventually I turned back and rode up the hill, encountering geese at every point trying to come down. After what seemed an age, I met Valli below the summit and rode round the side, still meeting geese on their way down. Finally I came upon a large group and brought them up. We ushered them safely into the cage: a catch of 180.

We made another cage at the front of the one that had been set so the whole cage was the shape of an hour glass. The catching and ringing went well, and we finished around 8.15 pm, leaving the net for a future catch. We had one and three quarter hours to reach camp in time for the radio schedule — and two rivers to cross. We raced back across Odkelsver — my pony leading at a fast triple, and we arrived just in time — but conditions were bad and we achieved no contact.

Next day we intended to get up early to make a botanical expedition up a mountain (Olafsfell) on the edge of the Hofsjökull, but the previous outing had been pretty exhausting, and neither Peter nor I appeared for breakfast before 11 o'clock, though poor Finnur had been waiting since 8. But we started out at 12.45 pm leaving James at camp with his bad ankle.

At Nautagi, where we had lunch, we took the temperature of the hot springs — the highest being 44°C — and Finnur collected some more botanical specimens. Many plants grow round the hot springs which are not to be found anywhere else. We spent nearly two hours there and then went on to Olafsfell. At the foot of a glacier Finnur found a rare plant which we had never seen before. It could hardly have looked more dull and insignificant, but it seemed fitting that it should be so, and Finnur was thrilled to have found it. Thus the Olafsfell expedition was entirely successful.

The far side of the mountain was very steep and the descent — even leading our ponies — was difficult; we were all wearing rubber boots and mine had quite smooth soles. We came half-way down on the glacier. It was soft enough for the ponies to sink in, but as I barely made an impression the only thing to do was to glissade: which I did.

Every now and then I was checked by my pony, whom I was still leading. I laughed so much that I nearly went on to the bottom, but finally managed the diagonal successfully and joined Peter and Valli at the point where the vegetation started again. We went on to the hot springs at the foot of Olafsfell. These springs were much hotter than the Nautagi ones, the temperature at source being 62°C. They were on the edge of the river bed and therefore swamped at times — so there wasn't much plantlife around them. But the water was deliciously hot — in fact just too hot to bear. It was a long ride home, but we arrived about 10.30 pm to find that James already had the steak and kidney on the boil.

We had been lucky with the weather, but the following day, when we rested, it rained from noon onwards. At about 5 o'clock I heard an aircraft engine, and we peeped out of the tent. There was a plane flying unusually low, and as it turned and flew in a circle we became very excited and rushed for our cameras. It made a wide sweep, lost height, and flew straight for the camp up wind. At about 100 feet it dropped its two bags, which landed five yards from our big tent — right in the middle of the three tents. Very nearly too accurate a drop! The pilot circled again and waved to us when we fell on the bags.

It was like opening Christmas stockings. There were films for Peter (urgently needed, as he had run out), letters, fruit juice, and newspapers. It was a wonderful piece of luck that they had dropped the bags on a day when we were all in camp. Flugfélag, one of the Icelandic airlines, carried out the operation for publicity purposes, so it was free.

Included in the drop was a letter confirming Peter's decree absolute — so divorce proceedings were now completed and we could be married on our return to Reykjavik.

The next day the weather was better, but the cold north wind blew and we went off for our final drives to the goose nets. On the first we drew blank, so we picked up the net and went on to the second, where we made a good catch of 114 — though some 26 were recaptures. Unfortunately we had failed to catch enough geese to finish all the rings, which was disappointing, as the ponies were due the following day for our return journey. We agreed, however, that we must make an effort to catch the last 46 required. On the way back my pony put her foot in a deep rut and went down. I turned a complete somersault over her head — but landed undamaged. One blessing about those small ponies was that there wasn't far to fall!

The next day (1 August) we had planned to go as far as the Hnifa, but on the way, and only fifteen minutes from camp, we came upon geese under a hill. We split up and hid behind the hill while Valli galloped round and drove the geese up. There was a large group of geese on top, and while the others stood round the flock I set the net and made a cage out of sight of the geese. Then the others walked the geese slowly down the ten yards or so and into the cage. There were 202, so we used all the rings as well as some wing clips. It was a wonderful success and we had been able to do it all by 2 pm and without going far. We still had about 20 goslings to mark when the relief party of two guides with twelve ponies appeared over the hill. So there we were with three guides and eighteen ponies. We went back to the camp for lunch and then started the slow, sad business of packing.

The morning of 2 August was the worst, but we had everything packed by noon. The guides then had to make up the packs for the ponies. With lunch in between, this took them until 3.30, when we finally departed — having made a bonfire of our rubbish and leaving behind eight ration boxes which couldn't be loaded on the ponies.

The return trip to Åssolfstadir took three days instead of the four on the way up: exhausting but worthwhile. We slept in the tents at Åssolfstadir and returned to Reykjavik by bus on Sunday 5 August.

The total number of Pinkfooted geese marked with rings or wing tags during our expedition to the tundra of the Thjorsaver in Central Iceland from 26 June to 2 August 1951 was 1,521 birds.

That was the end of the expedition, but two days later, on Tuesday 7 August, we embarked on a new adventure.

And so, at the end, you find my beginning.

Wedding group on the steps of Höfdi. Jack Greenway, who conducted the ceremony, is at the highest point of the photograph.